D1031315

The Art & Craft of Handmade Paper

Vance Studley

Dover Publications, Inc., New York

For Margo and Tyler

Vance Studley is currently a Professor of Art at California State University, Los Angeles and heads up the Advanced Typographic Studies Workshop at The Art Center College of Design in Pasadena, California.
He has won many awards for his work from The Type Directors Club of New York, The Art Directors Club of Los Angeles, The Art Directors Club of New York, The Society of Typographic Arts, and The Printing Institute of America. His work has appeared in *CA Magazine, Graphis, Print Magazine,* and *Typography 9*. He is the author of a number of books on the subject of crafts, lettering and typography.

Published in Canada by General Publishing Company, Ltd., 30 Lesmill Road, Don Mills, Toronto, Ontario.
Published in the United Kingdom by Constable and Company, Ltd.

This Dover edition, first published in 1990, is an unabridged, slightly corrected republication of the work originally published by Van Nostrand Reinhold Company, New York, in 1977.

Manufactured in the United States of America
Dover Publications, Inc., 31 East 2nd Street, Mineola, N.Y. 11501

Library of Congress Cataloging-in-Publication Data

Studley, Vance.
The art & craft of handmade paper / Vance Studley.
p. cm.
Includes bibliographical references.
ISBN 0-486-26421-1
1. Paper, Handmade. I. Title. II. Title: Art and craft of handmade paper.
TS1109.S83 1990
676′.22—dc20 90-34057
CIP

Contents

Preface 6

Introduction 8

1. **Handmade Paper 10**
   - The Spread of Papermaking 11
   - Ts'ai Lun 17
   - Growth of the Papermaking Process 18
   - Papermaking Today 26
2. **The Method of Making Paper 30**
3. **Materials, Tools, and Equipment 32**
   - Materials 32
   - Tools 36
   - Equipment 37
   - The Mold and Deckle 41
   - Hollander Beater 50
4. **How to Make Paper 52**
   - Scissoring the Raw Material 54
   - Boiling the Pulp 56
   - Washing the Pulp 58
   - Beating the Pulp 59
   - Coloring the Pulp 64
   - Sizing 64
   - Blending the Contents 65
   - The Vat 66
   - Forming the Paper 68
   - Pressing, Drying, and Separation 71
   - Sizing 76
5. **Projects 78**
   - Pouring Pulp over a Three-Dimensional Form 78
   - Embedding Objects into Paper 80
   - Layering 82
   - Pouring Pulp into a Form (Casting) 87
   - A Portfolio 94
   - Paper for Printmaking and Drawing 96
   - Paper Collage 100

Glossary 106

Bibliography 108

Index 110

Preface

Papermaking is an artistic craft of great antiquity, and at the same time, a highly mechanized modern industry. The division between craft and industry is not so wide, however, as might at first be imagined. Indeed, it is interesting to note that the three main problems presented to the mass-production paper-maker of today with his automated lines of vast machinery are those that confronted the early Chinese papermaker. The first problem is to find and utilize those materials suitable for paper manufacture; the second, how to pulp and form paper with some degree of durability and performance; and the third, how to make paper that is pleasing to the touch and attractive to the eye.

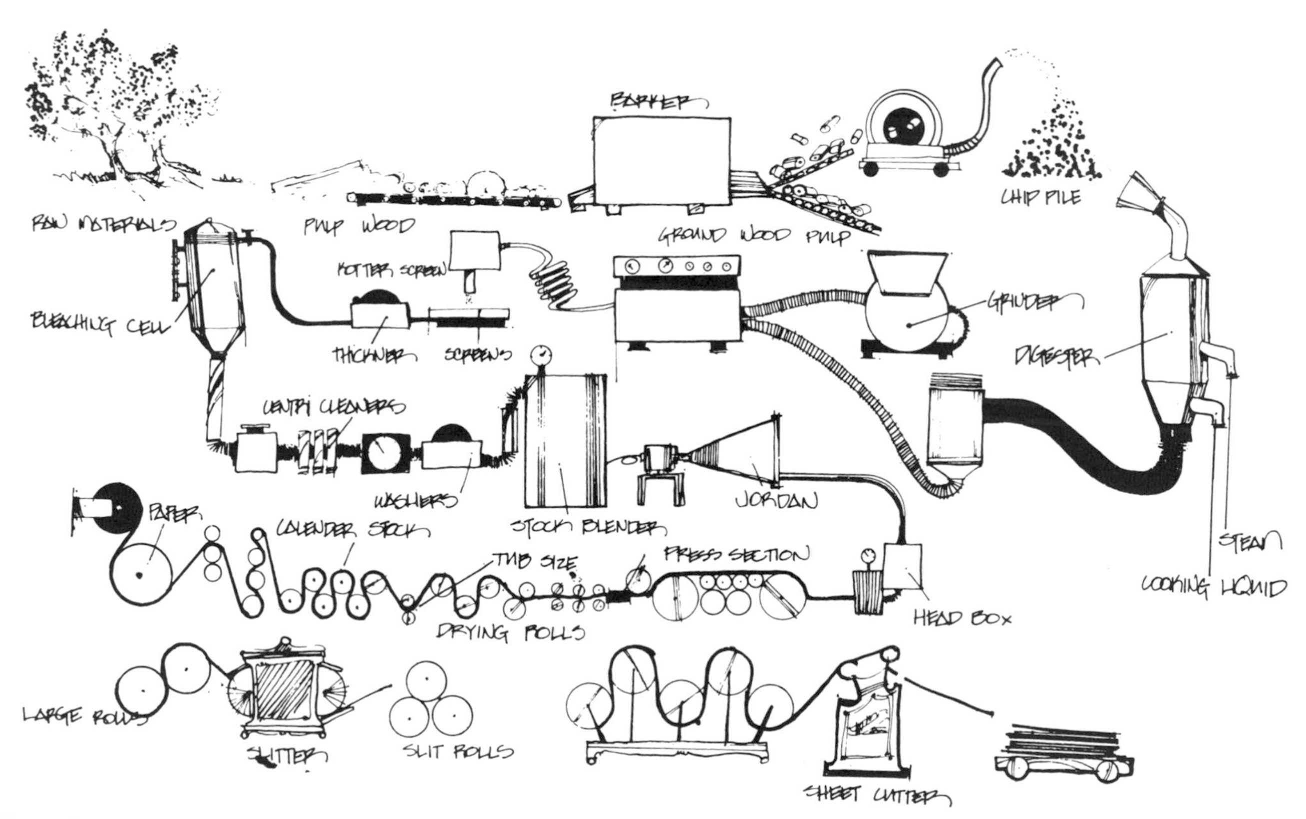

Illustration of how machine-made paper is fabricated using twentieth-century technology. It was in 1799 that Nicholas-Louis Robert invented the first paper machine. Today, commercial paper, in part, is made following the precepts of Robert's eighteenth-century invention.

These three problems are as fundamental now as they were during the formative years of the craft. They are normally solved by a careful analysis of all the variables. Thus, consideration of the size and composition of paper, thickness of paper, type of paper, its particular use and function, value of the image it enhances, particular requirements related to use, the expected life span of the paper, the number of sheets to be made, and factors concerned with the cost of the total papermaking process, must be considered.

Interest is always immediate when one is asked how paper is made. Interest grows further when one realizes that unusual papers can be made with little difficulty, and so made to assume new shapes and forms that have no identical counterpart. It is the author's intention to explain and illustrate simple papermaking procedures that are within the grasp of those wishing to learn how paper is formed, regardless of age group, and what enjoyment this simple act can bring.

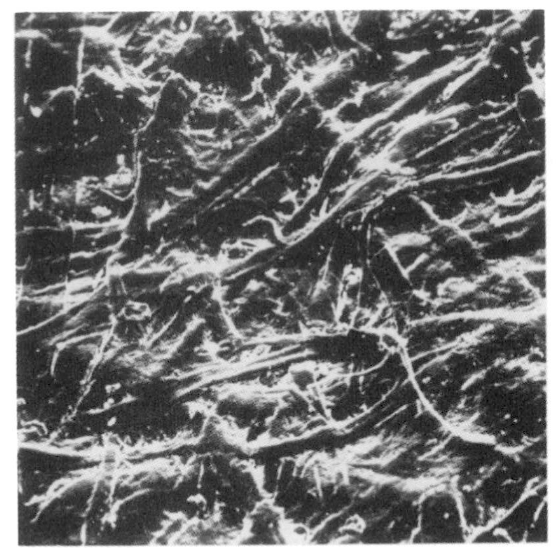

Scanning Electron Micrograph of paper currency magnified 500 times. (Courtesy of the Industrial Laboratory, Eastman Kodak Company)

Scanning Electron Micrograph of paper toweling magnified 240 times. (Courtesy of the Industrial Laboratory, Eastman Kodak Company)

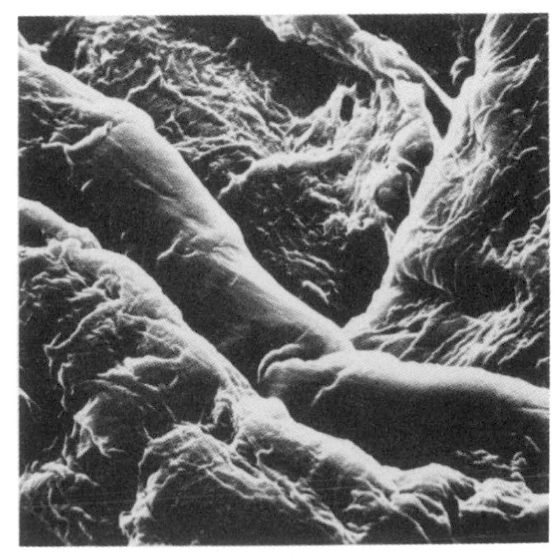

Scanning Electron Micrograph of paper magnified 4500 times. (Courtesy of the Industrial Laboratory, Eastman Kodak Company)

Introduction

This book on the art and craft of making paper by hand contains new approaches to the age old technique of transforming pulp into sheets of useful artist's paper. It offers an introduction to the tools, equipment, and materials needed together with sequential demonstrations of basic papermaking skills that are within the scope of an average school, college, studio, or home studio. The techniques to making paper are easy, the necessary tools are inexpensive, and quality results are had in a very short time. Artists of any age can benefit from this book. No previous experience is necessary. Teachers may use the book for classroom projects as a refreshing change from traditional modes of creative expression.

Part I introduces the origin and development of paper, its use in history, the people who made paper and continue to do so today, and why quality paper is ideal for creative use. This material has had an unusually colorful life in both the eastern and western parts of the world. Books, which today are so common, were once considered rare and precious objects, deserving the treatment and veneration accorded to great works of art.

Part II describes the necessary items needed to make small quantities of paper within the home, studio, or classroom. Tools may already be in the kitchen utensil drawer or in the shop. They may also be purchased from hardware or grocery stores offering shredders, choppers, spatulas, and other useful kitchen aids. The sections dealing with the construction of papermaking apparatus contain instructions and illustrations for making the basic paper forms (flat and multidimensional). More complex equipment for the student wishing to make greater quantities of paper or larger sheets are also shown with diagrams illustrating details of their construction.

Once the workspace has been organized and the papermaker has had the opportunity to make several sheets of assorted paper, then the projects described in Part III may be of further interest. Many useful ideas and directions are contained here. Paper for water-media can be made into various thicknesses, sizes, and weights depending on the individual use for each sheet. Paper may also be decorated by marbling techniques employing an assortment of inks, dyes, paints, and other coloring. Also described in this portion of the book is paper to be used for simple forms of printmaking (with or without a press) which will enliven the textural richness of the image.

Projects of particular interest to young artists are so noted in the introductory remarks preceding each section.

Although the initial results of early experimentation with the basic techniques and materials are usually almost unpredictable, such results, owing to the nature of the craft, are surprisingly inspiring and beautiful. Students can extend the various projects to create their own paper with individual character and texture. Once the new papermaker becomes familiar with a technique, he may expect to be more discriminating and selective, to achieve some degree of control over the results of his work, and to succeed in arranging colors, textures, weights, and shapes as he chooses. The opportunity to experiment with each project exists as long as the artist remains open to newly discovered results and further ideas.

Paper can be a luxurious material that is an end unto itself with no further need of decoration or embellishment with artist media. Such is the variety of finishes that can be accomplished with the blending of common ingredients. The student may wish to create a folio of papers or possibly to bind them by hand for permanency, thereby creating a "picture book" or journal of experiments.

Some time- and space-saving techniques arrived at through much trial and error in successfully making the papers described herein are offered. Therefore, it is recommended that the reader proceed through the book in the order in which it is presented, creating four or five papers from each section before proceeding to the next. Each successive section does not become more difficult. However, each lesson serves as a basis for further experimentation requiring an understanding of the rudiments of pulping, forming, drying and sizing.

Once you get involved in the making of paper by hand you will find yourself taking notice of paper around you, appreciating its composition, texture and its versatility. Whatever its function, paper makes a valuable contribution to our sensory experience.

A close examination of Degas' pastel drawings will reveal the use of textured papers with many earthen tonalities. Picasso's use of a variety of intriguing papers is no more evident than in the collage constructions of his work in Cubism. Watercolors by John Marin and Paul Klee show the use of paper as an integral part of the image. Buckminster Fuller has advocated the construction of a geodesic dome of paper and bamboo.

As can be seen, ideas are limitless. Fool around with your paper. Draw on it. Paint on it. Print on it. Fold and score it. Pour it. Or, do nothing but sit back and admire it for its own inherent beauty.

1. Handmade Paper

The importance of paper in our lives today is inestimable. It is the bearer of thoughts, recorder of knowledge, promoter of commerce; it is the surface on which lie many of the world's greatest artistic treasures. Oddly, little has been written about the ways in which the artist, the craftsman, and the student can make their own paper. Quality paper yields unsuspected richness of texture and possesses a tactility which enhances the images made on its surface. A piece of paper is not just a piece of paper, but becomes an essential ingredient in the creative process and can give life to the artist's mark. For all the differences there are between kinds of papers—rag paper, wood, watercolor, newsprint, etc.—they all serve the user for various needs.

Today's artist is no longer satisfied with paper as merely a surface to hold an image. He wishes to extend the use of paper and pulp by utilizing flexible form and responsiveness as integral parts of what artists do best with media: give shape and unity to ideas.

The lithographs of Daumier, the etchings of Rembrandt, the velvet-like aquatints of Goya, and the pastels of Degas were executed on papers of incomparable quality which survive today as testimony to the handcrafted expertise of the European papermaker. Fine paper is an inspiration to artists whether it be in sheet form or dealt with sculpturally as liquid paper. It can be formed, poured, shaped, embedded, layered, molded, and colored. Paper can be made from cotton rags, celery stalks, iris, gladiola, straw, wheat, bamboo, cattail, hemp, potatoes, reeds, beans, linden leaves, burdock stalks, St. John's wort, horse chestnut, hollyhock, marsh meadow, thistles, and blue grass. The list grows with each experiment of the papermaker.

The requirements are simple. The reader will need several basic items: a workspace, raw materials (from your garden or grocery store), several hand tools, one or two hand appliances, and a flat drying area.

Oracle bone from a tortoise. The oracle statement concerns a harvest and its divination, ca. 1300 B.C.

Inscription engraved on ox bone, thirteenth-century B.C. Ink rubbing.

It should be remembered that the first few attempts at forming pulp into paper will take practice and the initial sheets may look crude, uneven and quite unlike paper you are accustomed to seeing and using. Save these first sheets; they make excellent materials for collage, mobiles, and papier-mâché, and can always be repulped at a future time for new sheets or as an additive to other pulp for exciting mixtures of color and composition. It is from this starting point, from these unexperienced beginnings, that skill can grow, transforming the crude to the polished originality that is within each student, artist, and craftsman.

THE SPREAD OF PAPERMAKING

Paper serves us so constantly in our daily lives that it is difficult to imagine a world without its benefits. On every side we are confronted with books, magazines, newspapers, and posters. Most of our knowledge of the past comes to us on paper; most of our current information is printed on paper; and the record of what we do and say and think today will be made available to the future on paper. It is difficult to conceive that in ancient times there was no substance of this kind. Tracing the various materials in use prior to the advent of paper makes us realize how impossible it would be for modern civilization to endure, even momentarily, the total lack of this indispensable material.

Written language is an essential condition for the development of communication. Before written language, there was only spoken language, often guttural in sound. Primitive people passed on their knowledge by word of mouth, depending on memory to accurately record myth, event, and story. The Chinese, like the people of other lands, had used the method of tying knots. Different knots—large or small, single or multiple, loose or tight, in various colors—indicated many things. Eventually, written symbols evolved as communication.

Picture drawing or pictograms was another way by which primitive people expressed themselves. Chinese script, especially, was a product of pictographs from which the Chinese written language evolved.

In China the earliest writing had no fixed form. It was done on the protective covering of certain animals such as tortoises, on mammals, skin, bones, and later, on bronze, stone, bamboo, wood, silk, and finally, paper. The invention of paper made the book possible. It was the surface on which man was to create his images of religious belief in which the artist was so instrumental.

Oracle Bones

Oracle bones date back to the sixteenth century B.C. Scholars learned that the marks on these bones were primitive forms of ancient writing recording the events of that time. The inscriptions contained records of questions asked of the gods or divine ancestors. Before going into a battle or offering sacrifices, or in the event of sickness, the Chinese subjected the shells or bones to the extreme heat of fire. The cracks which resulted on cooling were interpreted by the wise men.

On the shells or bones were recorded the reasons for seeking guidance and the divine instructions as interpreted. These were valuable records which were kept for later verification and as a result, were preserved carefully. This preservation has provided today's scholar with much information about the early lifestyle of the Chinese.

Stone

Stone was most likely the first material upon which effigies and, later, characters and letters were graven in other parts of the world. Through this medium of expression an inordinate number of historical records have been made available to the modern world. With sharp carving tools Egyptians carved hieroglyphs in monuments of stone, called obelisks. These four-sided shafts terminate in an abrupt pyramid. One such ancient obelisk stands today in Central Park in New York City.

Greater seal script inscription on stone, China, fifth century B.C. Ink rubbing.

Clay Bricks and Tablets

The Chaldeans of ancient Babylonia impressed characters with incising tools of bone into clay bricks or tablets of various sizes. This is known as cuneiform, meaning wedge, because the marks made in the clay were wedge-shaped. The clay tablets were then baked until hard. These tablets were transmitted from one person to another, just as letters and accounts written upon paper are exchanged today. Ashurbanipal, the Assyrian ruler and general, formed a great library of clay tablets, setting scholars to collecting literary, historical, religious, and scientific works, and to making translations and editions of valuable thought.

Metals

The use of such metals as brass, copper, bronze, and lead was not unknown to early civilization. In the Bible, reference is made to the use of lead for permanent writings. Other metals were used in preserving treaties, laws, and alliances. The Romans used bronze in recording their memorials, and on the field of battle Roman soldiers engraved their wills on their metal buckles or on the scabbards of their swords. Being a favorite possession, bronze articles were inscribed at first with names or other symbols to show ownership, later with writing to commemorate an occasion or explain the reason for making the bronze objects, or to describe their use, or to record the names of their makers.

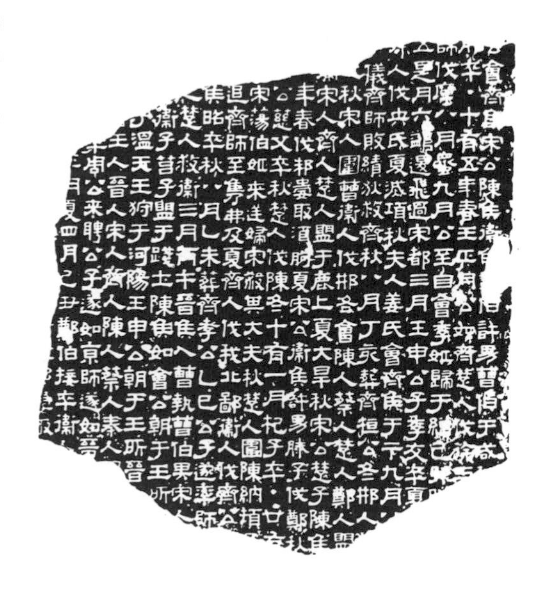

Stone inscription from China, engraved on both sides, A.D. 178–185. Ink rubbing.

Clay tablet showing use of cuneiform. Note the incised lines created with wedge-shaped tools. The text registers animals brought in by different individuals. (Courtesy of Muir Dawson)

Enlarged portion of cuneiform inscription on clay tablet.

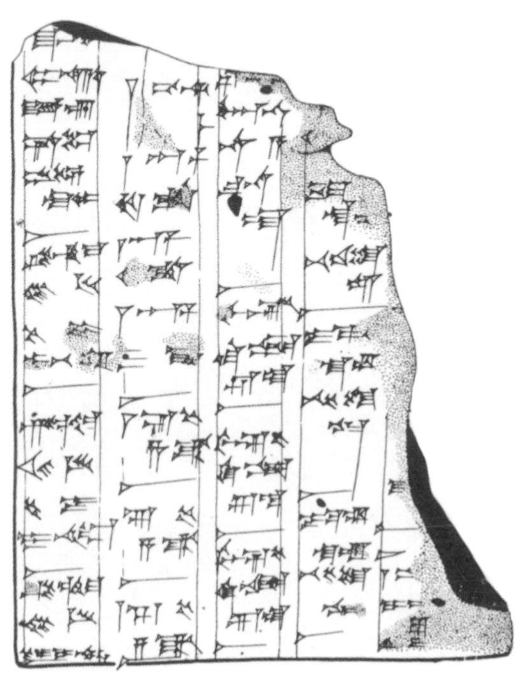

Linear description of cuneiform inscription showing various shapes achieved with assorted tools.

Wood

Large books constructed of pieces of wood, were in use before the time of Homer (ninth century B.C.). The main material came from the box and citron trees. Each section of wood was usually covered with a thin veneer of wax, chalk, or plaster, and letters or marks were scratched onto the coating with a metal or bone stylus. This technique permitted erasure by recoating the wooden boards. The separate boards were fastened together with leather thongs and thereby composed a book, called a codex. The table-book remained in use until the fourteenth century A.D. In the Orient, characters were inscribed into strips of dried bamboo which were strung together to form a bundle. This object was cumbersome and difficult to store. Each time a book was used the string had to be untied and re-tied. The ancient Chinese did not number the strips and confusion occurred if a string snapped and the bamboo strips were disarranged. Replacing them in their proper sequence was, often times, never done.

Vegetation

Writing on palm and other species of leaves was practiced from the earliest periods in Rome and Near Eastern countries. From the large palmyra leaves, strips of varied length and about two inches in width were cut. A metal stylus was used to make indentations on the leaf and these scratches were filled with a dark paint derived from the carbon in ashes making the letters clear and distinct. Each leaf was pierced with two holes, and the leaves were strung together with cords to make a book. From the use of leaves of assorted trees for the making of books in ancient times the word "leaf" is now used to mean a part of a book.

Close-up of palm leaf inscription showing incised lines filled with dark pigment.

Palm leaf book in closed position. The leaves of the book are strung together with cords. Pages were not numbered and often became disarranged. (Courtesy of Muir Dawson)

Bark of Trees

The bark of many trees has been used as a writing material in almost every period and region. The ancient Latins used the inner bark, known as liber; in time the term liber denoted a book itself, and from it our word "library" originates. The American Indians wrote their language of symbols with wooden sticks and liquid paint upon the hammered bark of the white birch trees of northern America. The native aborigines of Central and South America, including Mexico, have, on many occasions, made a type of paper by beating the inner bark of moraceous trees. History does not show that the native aborigines of what is now the United States ever made paper of any form or type.

Parchment

Parchment, a sheetlike material made from animal skin, occupies a unique and highly regarded place in the history of paper evolution. It evokes a sense of antiquity and is associated with quality that is practically without peer. Parchment is durable and has for many hundreds of years withstood the rigours of age and continuous use. Parchment has been the bearer of valuable records and stories from the age of classical Greece to medieval times, and many ancient parchments survive to the present day as witness to this exceptionally useful material. The word "parchment" derives from Pergamum, an ancient city of Mysia in Asia Minor. Scholars feel that parchment was probably in use as early as 1500 B.C., however, it was not considered to be a common writing surface until much later, most likely around 200 B.C.

Near Eastern calligraphy on parchment. (Collection of the Author)

The skin of animals has proven to be most difficult to prepare and has its attendant problems: soft versus hard, porosity, and surface acceptance to various media. Animal skin is tough and is capable of much manipulation and treatment at the tanner's hand. It may be colored, polished, flexed, and decorated by embossing, incising, punching, and stitching. It ages slowly in use and when colored, scraped, incised, or simply stored unused, retains a beauty of its own. True parchment, unlike leather, is made from the split skin of the sheep. The grain or wool side of the animal's skin is made into skiver, that is, material suitable for use in bookbinding. The flesh or lining side of the skin is converted into parchment of the finest quality.

Enlargement of parchment surface.

Vellum

Vellum is made from calfskin, goatskin, or lambskin and is usually made from the entire skin. Vellum often may be distinguished from parchment by the grain and hair marks which produce a somewhat irregular surface. Parchment is more consistent in appearance and does not possess these elusive characteristics. Parchment and vellum must be scraped, rubbed with lime, and stretched so the skin takes on a uniform appearance. It is then sanded with fine pumice making it an ideal surface for writing and calligraphy. Scribes throughout the ages have been very selective with the skin to insure uniform color and surface quality in bindings requiring many pages. Parchment remained in continuous use through the Renaissance, and it is said that to produce a single copy of the Gutenberg Bible required the skins of three hundred sheep. Parchment and vellum are still in demand and use today owing to the nature of fine diplomas, certificates, patents of nobility, etc. Calligraphers find these materials to be ideally suited to the extremely fine detail required of good workmanship. If paper had not been invented, expensive parchment would not allow for the extensive written communication we have enjoyed for many hundreds of years.

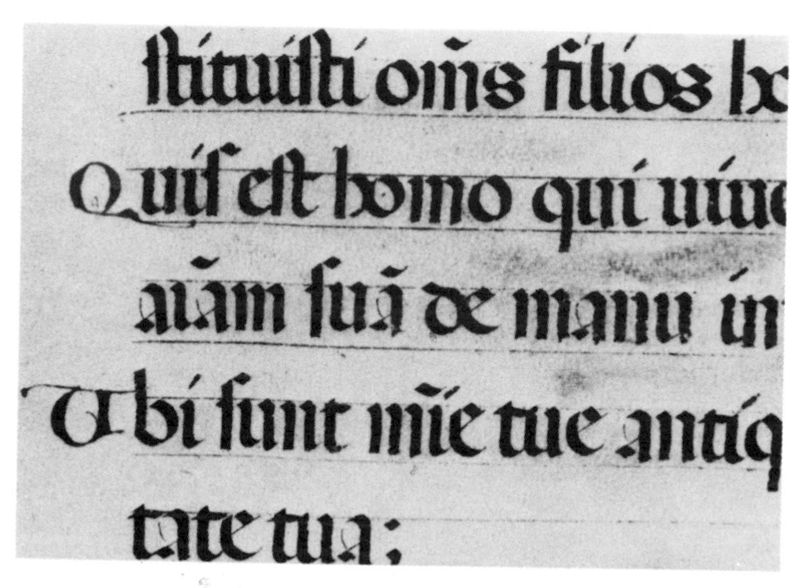

Vellum with sixteenth-century writing. Note the porous surface which has been scraped for a smoother surface. (Collection of the Author)

Papyrus

The first successful attempt to manufacture an article resembling modern paper, so far as is known, was made in Egypt at a very remote time. An aquatic plant known as papyrus, afforded the material. Papyrus is a strikingly attractive plant, the stem of which grows from ten to twenty-five feet high. The stem is triangular in cross section and around its base grow several short-fibered leaves. It is smooth without any knots and tapers gently towards its flowery cluster, which is large, delicate, and tassel-shaped. The plant grows profusely in the stagnant shallows of lakes and rivers in many parts of Africa. It is from the thin coats or pellicles of the papyrus that papyrus paper was made. These were separated by means of a sharp, long pin, or pointed mussel shell, and spread on a table with a thin stratum of water in the form the size of the sheets required. On the first layer of these slips, a second was placed transversely to form a sheet of desired thickness, which, after being pressed and dried in the sun, was polished with a shell or other smooth, hard substance. Twenty-two sheets were the most that could be separated from one stalk, and those nearest the pith or center made the finest paper. The commerce of Egyptian paper was flourishing in the third, and continued until the fifth century B.C.

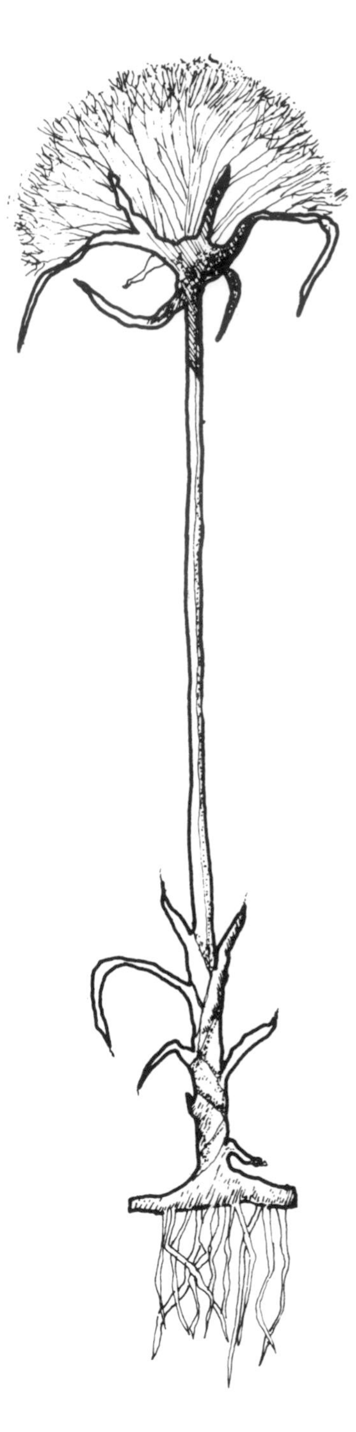

The papyrus plant.

The growing use of animal skin, and the geographical changes of the Nile region contributed to the demise of papyrus. Cultivation became difficult and papyrus declined rapidly. The words "paper," "papier," "papel," etc. are derived from the Greek word *papyros. Biblios* was the Greek term used to denote the inner fiber of the papyrus plant, and writings on sheets of papyrus were known as *biblia.*

There are many other flexible writing and drawing materials that may be compared with papyrus: the Amatl bark papers of the Aztec and Mayan peoples, and the so-called "rice paper" of Formosa. Rice paper, which is a misnomer, is a thin substance cut spirally from the inner lining of a tree indigenous to the country, and has no relation to rice, its derivatives, or to true paper. It is used in China primarily for sumi painting and calligraphy.

Paper

Paper is a material made in the form of thin sheets from rags, straw, bark, wood, hemp, or other vegetable material. To be classified as true paper the thin sheets must be made from fiber that has been macerated until each individual filament is a separate unit. The fibers are intermixed with water and, by the use of a fine mesh screen, lifted from the water in the form of a thin stratum. The water drains through the fine openings of the mesh or screen, leaving a sheet of knitted, clotted fiber upon the screen's surface. This thin layer of intertwined fiber is considered to be paper.

TS'AI LUN

This is the manner in which the Chinese court official, Ts'ai Lun, formed the first paper in the year A.D. 105. To this day the precepts and methods of Ts'ai Lun have remained much the same with large machines and sophisticated equipment employing the same principle. The very formation of paper fiber has undergone no change in almost two thousand years.

Ts'ai Lun conceived the idea of making paper from old rags, hemp, tree bark, and fish nets, and reported his invention to Emperor Ho-ti. He was acknowledged by the Emperor for his efforts and became known throughout China as the patron of papermaking.

There is some dispute as to whether Ts'ai Lun was indeed the true inventor of paper. Scholars feel that a kind of paper made from silk may have preceded his product. The credit of original invention of a fused paper and silk product can be attributed to the people who first used silk waste to make paper, and the time was before A.D. 105. It was known that in the year 12 B.C. paper was already used for wrapping medicine in the imperial court. The fact that Ts'ai Lun retained the old name *chih*, a partial word for silk, instead of using a new name was further proof that paper was not an entirely new product but only an improvement on something which already existed.

GROWTH OF THE PAPERMAKING PROCESS

The art of fabricating paper was zealously guarded by the Chinese. It was not until A.D. 600 that papermaking reached Korea, and about fifteen years later, Japan. In the eighth century the craft reached Samarkand. From there paper-making traveled to Egypt and then to Morocco. Not long after, the industry reached Europe. In the twelfth century the Spanish were creating paper in Valencia. From there it quickly spread to France, where the first mill was set up in Troyes. Finally the craft of papermaking reached England in the late fifteenth century.

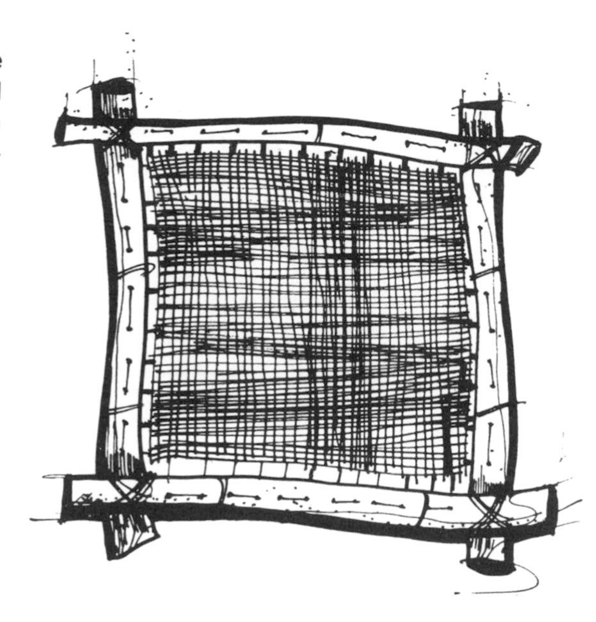

Crude mold from China. This mold is similar to the ones most likely used by Ts'ai Lun in his first papermaking efforts. Made of bamboo and grass, this mold requires the pulp to be poured upon the surface. The mold and paper were then left outdoors to dry.

Early papermaking technique of drawing pulp from the vat with the aid of weighted pulleys.

Italy
Meanwhile paper was being made in Italian provinces. One of the first documented references to papermaking in Italy is a late thirteenth-century reference to mills at Fabriano, where paper is still made today. The paper mills at Fabriano remained the most important, producing an unusually fine quality rag paper sized with animal glue. This type of manufacture was very popular with the scribes of the day. The surface was smooth and excellent for writing and drawing. The use of animal sizing led to a greater acceptance of paper as a substitute for animal skin (parchment, vellum) and the technique was quickly adopted by papermaking mills all over Europe.

Early papermaking mill in Italy showing various phases of the hand-papermaking process. (Collection of Pictures of Old and New Papermaking for Handmade Paper)

France, Germany, and The Netherlands
In France there is a legend that in 1147 a French citizen, Jean Montgolfier, was taken prisoner during the Second Crusade and sentenced to labor for three years in a Damascus papermill. There he was able to learn enough of the craft to set up his own papermill after he returned to France. During the next two hundred years, papermills flourished in France and for many decades these mills provided the Low Countries with fine paper for their express needs.

There exists evidence that papermills were producing paper goods in Cologne and Mainz, Germany in 1320. It is in the diary of Nuremberg papermaker Ulman Stromer that documented reference is made of an actual mill which Stromer built and equipped in 1390. Stromer's mill, located near the city of Nuremberg, is the first papermill to be pictured in a published book: Hartmann Schedel's *Nuremberg Chronicle* of 1493.

Papermaking became an established craft in The Netherlands during the late sixteenth century. The Eighty Years War was in progress then and ties with the papermaking centers of France had been severed the year before the capture of Antwerp by the Spaniards. This created a migration northward from Antwerp to Amsterdam. By the end of the century Amsterdam became a melting pot of papermakers of many nationalities and the city grew into an international trade center. Just prior to this period the great Dutch artists Rembrandt, Vermeer, and Frans Hals created many of their masterful works in and around Amsterdam for many guild merchants. The guilds oversaw the flourishing industrial trades that came about as a result of increased commerce. With the invention of the Hollander beater, a device used to macerate and prepare paper pulp, in the latter part of the seventeenth century, papermaking became a major industry, and Holland became known for its fine, white papers.

The papermill at Nuremberg established by Ulman Stromer in 1390.
(Hartman Schedel, *Liber Chronicarum*, Nuremberg, 1493)

The steps of papermaking: forming, pressing, and drying. Early woodcut.

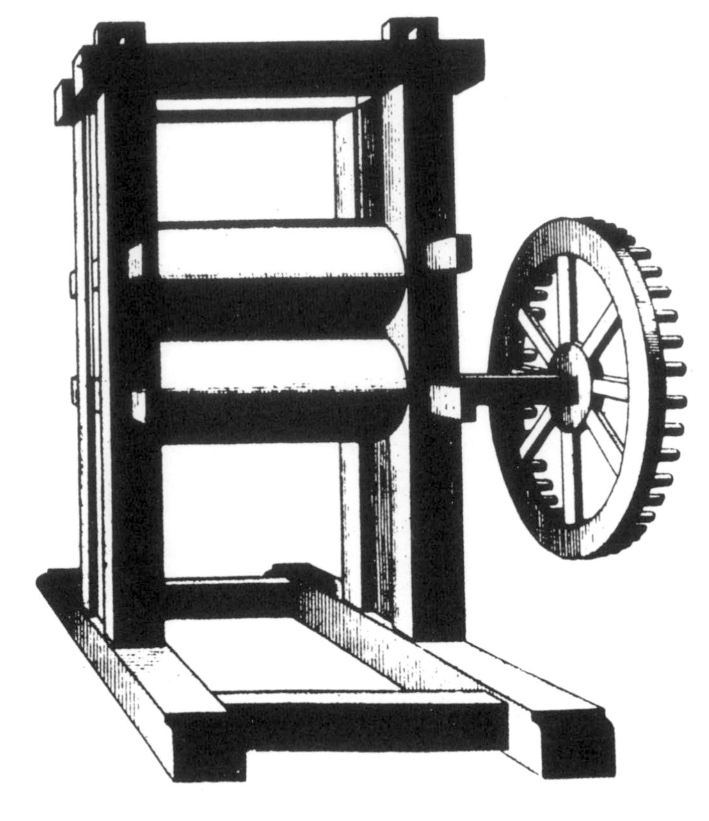

An early calender roll used to impart a smooth finish to the new sheet. (J. J. Le Francais de Lalande, *Art de Faire le Papier*, Paris, 1761)

A sixteenth-century machine for macerating papermaking material.

An early series of engravings showing various stages of the papermaking process from beating to drying.

Illustration of printing and its technique on handmade paper. (Ernest Lehrner, *Signs, Symbols and Signets*, World, 1951)

Today, the Van Gelder Paper Company still operates a wind-driven paper mill in Holland. The mill is located in Westzaan and was built in 1692. The paper made at De Schoolmeester, so-named because a minor shareholder in the original mill was a schoolmaster and later a notary public, is crude and very coarse. Unsorted rags and rope and raw canal water are the basic ingredients. This paper-making *au naturel* yields sheets with a variety of plants, insects, parts of fishes, and other unusual and fascinating odds and ends. The mill makes three different papers including moss-paper which contains up to ninety-percent moss. Peat-moss is gathered along the banks of brooks which babble through the sandy heart of The Netherlands. Fibers, water, wind power, and the skill of the miller/paper-maker, these therefore are the ingredients that go into the making of De Schoolmeester's paper.

A portion of paper and printing on paper from De Schoolmeester in Holland. The paper is a combination of rough fiber and moss. (Courtesy of Henk Voorn, Haarlem, The Netherlands)

A wind-driven papermill similar to De School-meester in Holland. This papermill yields paper of unusual character resulting from the use of crude, coarse pulp from the local water banks. (Leendert Van Natrus, Jacob Polly and Cornelius van Vuuren, *Groot Volkomen Mooleboek*. Amsterdam, 1734)

Latin America

Of the thousands of items of early books, maps, records, official messages, and tribute rolls that exist from the European world while papermaking was in its infancy, there are extant but a few items of Mayan and Aztec manuscripts and paper fragments. The Mayas and the Aztecs made a paperlike material by beating the bark of fig (Ficus) and mulberry (Morus) trees. The Mayas called their bark paper *huun* and used it in making their hieroglyphic charts. The Aztecs made a substance called *amatl* by boiling bark strips into a heavy, glutinous ingredient that was then beaten and scraped into sheets. This ancient approach to papermaking is still practiced among the Otomi Indians of southern Mexico. Probably, the first mill of sorts in Mexico devoted to paper manufacture was established near Mexico City by Spaniards who were given a concession by the Spanish Crown to manufacture paper in New Spain, using material they had found. The remains of the mill no longer exist.

Illustration from Latin American Codex Monteleone, sixteenth century. On amatl, a native paper, fashioned from bark. (The original is in the Harkness Collection, Manuscript Division, Library of Congress)

The United States

Papermaking as an industry was introduced into the American Colonies largely through the efforts of William Rittenhouse, a German immigrant. In 1690 Rittenhouse and a group of influential citizens from Germantown built the first American papermill near a German settlement in Pennsylvania. The mill was swept away by high water in 1700, but Rittenhouse and his son Klaus built another mill near the original site, and it was in operation in 1702. No authoritative picture exists of the original mill or of the second structure.

The mills of this period were crude, dank, and poorly lit. Working conditions were such that it required a strong breed of worker to carry out the laborious task of paper manufacture. Equipment was usually handmade of hard woods and largely invented or modified to suit the needs of each mill. A Philadelphia antiquarian wrote: "The Rittenhouses, one or more of them, have continued as papermakers to this day. The first stone mill is attached to the present mill and covers a large press with a screw, made of cherry, about eighteen inches in diameter in the eye. The press verily appears to be an ancient affair. The mill is not at present

Enlarged watermark of the papermaker, Klaus Rittenhouse. (Broadside printed by Andrew Bradford, Philadelphia, 1726)

employed by a Rittenhouse. It is rented as a paper mill; and the present papermaker Nicholas Rittenhouse employs another mill in the vicinity . . . . Jacob Rittenhouse made the large wooden screw, which was about the year 1755 for his uncle William. Jacob, the millwright, also invented, about the year 1760, the slanting plates for grinding rags; as straight ones only, with much delay of grinding, were in use before this discovery . . . . Jacob is still living, he is now 86 years of age, is blind, but retains an excellent memory unimpaired."

The Ivy Mill, founded in 1729, and constructed by Thomas Willcox and others on Chester Creek in Pennsylvania. (Henry G. Ashmead, *History of Delaware, Pennsylvania*, Philadelphia, 1884, p. 492)

An early American papermaking machine. This machine was capable of forming paper and passing it along to the felts. From this point the machine would separate the felts from the paper and forward the sheet to the post. (Robert Clapperton, *The Papermaking Machine*, Oxford, 1965)

Mills began to spread throughout Pennsylvania. One mill supplied Benjamin Franklin with large quantities of paper for his printing and publishing activities. This was the Ivy Mill which proved to be a training ground for fledgling papermakers who in turn went on to establish their own mills in other Colonies. By the twentieth century, practically nothing remained of the original structures that dotted the eastern states, save for a few foundation boards and several small buildings which are of dubious historical accuracy or importance.

It would not be until the work of Dard Hunter, this country's eminent papermaker and historian, that this age old craft was revived. It is with Dard Hunter that the lore of paper forming by hand becomes a significant process which deserves a much closer examination. Hunter has written widely on the subject, and his books have been a valuable source of guidance to artists and craftsmen interested in utilizing paper pulp in their visual statements and ideas. Refer to the bibliography for further reading of the works of Hunter and his Mountain House Press in Chillicothe, Ohio.

PAPERMAKING TODAY

The above survey of the spread of papermaking throughout the world dealt with paper only as a surface material to be used as an adjunct to other forms of communication. But what of the nature and possibilities of manipulating paper pulp as an art form? Papermaking has gone well beyond its infancy and its attractiveness is undeniable. A closer look at paper provides an elusive experience. When one sees paper it is usually blank, white, or decorated, often without character or presence. It can be, and often is, a completely anonymous material which is used with little sensitivity or thought. To the artist, this presents an enigma: what does one do with a blameless material; how does one improve it?

The appearance of paper after it has been poured over a three-dimensional form, separated, and then dried. Paper dealt with in this manner is capable of great visual richness and texture.

Paper sculpture achieved by the fusion of natural and man-made fiber. The creative use of pulp lends itself to new and unusual forms. (Artist, Tom Mossman)

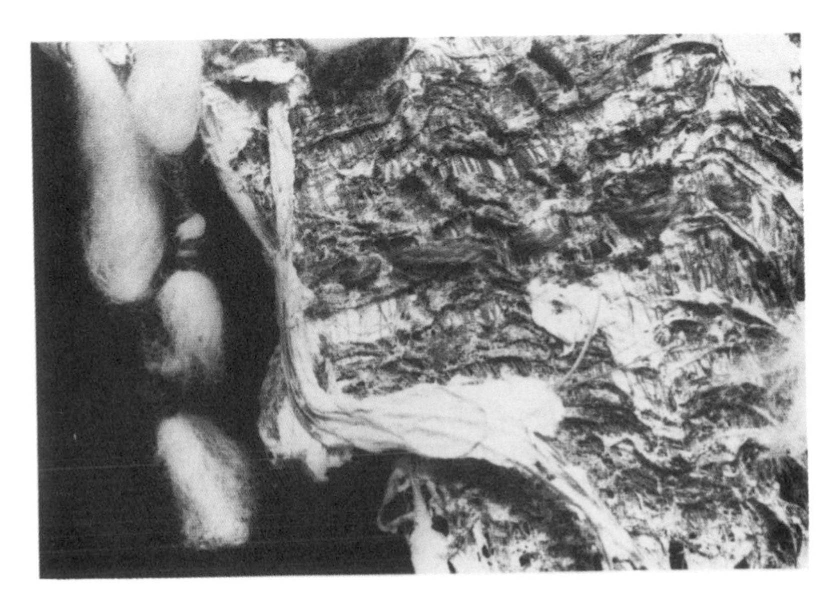

A closer look at the work.

The surface of paper encloses infinite space. Puncture it and one disrupts the interwoven fiber's chain, its mettle disturbed. Nonetheless, a mark is made and the beauty of the material becomes alive with various possibilities. Tear it, score it, fold and glue it, mark on it, or sail it and paper is airborne. It becomes something. Out of this involvement with paper expressiveness, one can see a transformation of paper into a material for creative purposes.

Attitude

The reader should make paper while maintaining a dialogue with it. Rapport with the pulp, with the act of forming it, will bring to the paper a harmony and texture that is unobtainable in machine paper. Its inherent beauty will speak to the maker's mark and allow it to grow with each sheet. The making of paper should be done rhythmically and consistently.

In Oriental homes the shoji panels, with a single sheet of paper, demarcate and create spaces for quiet living. The paper retains heat yet is porous enough to allow the circulation of gentle air. It is intended to obscure our vision; it permits a soft, warm light to fill the interior and gently illuminate the people there. Yet it is structurally weak and can be the flimsiest of materials.

Nothing can be made well, functionally or aesthetically or both, without some degree of organization, that is, design or some kind of motivating impulse. In the making of paper by hand, no one sheet must be like another, for the objective in the artistry of papermaking is not to make the same paper time and again with no surface or textural variables. This becomes rote and monotonous.

The artist must be capable of detecting the spirit of the paper stuff to create a new experience from it. Paper should reveal craftsmanship and creatively direct the flaws and intangibles to support the message, not to suffer as a result of it. It is difficult to make mistakes in forming sheets of paper for artistic use. A hand thrown ceramic pot does not have to hold tea to become viable. But it should show sensitivity, thoughtfulness, and the elements of design that make it unique and aesthetically pleasing. This is what papermaking and paper use by artists can be about: the transmutation of raw material into new forms that say something about the artist and his ideas in a medium that is best suited for his expression.

Watermarks

A watermark is a translucent marking made in paper while it is still wet for purposes of identification of the paper. The illustration shows wire figures and letters, oftentimes made from a continuous strand of brass. After the wire figure is formed it is sewn onto the sieve of the mold and thereby displaces the thin layer of pulp as it spreads over the surface. When the sheet is finished and dry, a transparent image of these wire figures becomes visible.

The watermark and its relationship to the mold wire.

Watermarks are strictly an occidental feature since no watermark has ever been found on handmade papers from the Orient or Near Eastern countries. To date, the oldest known watermarks are found on papers made in Italy at the turn of the thirteenth century.

No one knows precisely how the watermark came about. It has been suggested that an Italian papermaker accidently overlooked a small piece of wire on the sieve and soon discovered that the paper was thinner at this place on the finished sheet.

Throughout history to the present, watermarks depicting mystic, religious, and other worldly symbols have become a standard feature in papers of the western world. The watermark may be likened to a signature or trademark of individuals or guilds involved in the papermaking process and is associated with quality revealing the papermaker's pride in craftsmanship.

Various watermarks of European design, fifteenth-eighteenth century.

2. The Method of Making Paper

Paper is made by the layering of short vegetable fibers to form sheets. The hunt for interesting raw materials can be an adventure. Materials such as jute, artichoke, straw, iris, and bamboo are capable of being reduced to a state suitable for paper forming. The fibers must first be shortened and separated into fine fibrils or strands similar to bamboo after it has been crushed or beaten with a mallet. Once the raw material has been selected it is deposited in a large pot or cauldron to which is added a small amount of caustic solution and water to further remove impurities from the material. After two to five hours of simmering the liquid is drained off, and the above process is repeated once again. If need be, a third boiling may be performed.

By now the vegetable material is a pulpy mass with a somewhat slick and fatty appearance. Much of the unwanted dirt and lignin (an organic substance which acts as a binder for the plant material) has been separated from the crude pulp. The pulp is mixed with clean water and placed into the blender where a bladed apparatus will shred the mass and draw it out into a semiliquid, fibrous state.

The pulp is then poured into a tub or "vat" and more water is added to give proper consistency for casting or forming the sheet on the hand mold.

The creamy pulp, scooped up by the mold with the deckle frame attached, is given a two-way rolling shake to "throw off the wave." This intertwines and crisscrosses the short fibers. The fibers are now drawn together by the force of the strong suction created as the mold is brought to the surface. A thin stratum of pulp has now been formed on the mold. The mold and clinging matted sheet are allowed to drain briefly. The deckle frame is removed from the mold, revealing a slightly irregular edge around the periphery of the sheet, and the latter is "couched" or pressed face down onto the awaiting dampened felt. The mold is lifted, leaving the newly formed sheet on the felt.

The process is repeated until the felts and paper, which have been stacked upon each other, are built into a pile or "post." The post is then inserted between two flat boards or placed into a press, then squeezed to remove excess water. Sufficient pressure is applied so that the fibers are matted well enough to allow removal of the paper from the felts. The paper can then be placed elsewhere to dry.

The sheets must dry slowly, usually in the open air or in a moderately warm room to prevent ripples. The felts are then washed and prepared for their next use.

Once the sheets have dried to a normal moisture content, they can be sized. This is usually accomplished by immersing the sheets in a shallow tray of prepared gelatin sizing, removing them, and then repeating the pressing and drying procedure.

At this time the papermaker may decide to impart to the newly formed sheets a glazed finish suitable for drawing and calligraphy. The process is much like the ferrotyping technique used by photographers to give prints a glossy surface. The sheets are placed between two smooth tin, chrome, or zinc plates and placed into a press to give the required gloss to the paper. The press is then tightened. This technique will further mat the material and will create what is known as a pressed finish.

3. Materials, Tools, and Equipment

The necessary tools for making small amounts of paper by hand will vary to the extent to which the maker intends to carry his work. Disarmingly beautiful paper can be achieved with little equipment, with the procedures easily adapted to classroom needs. The lack of expensive or cumbersome equipment should in no way prevent the budding papermaker from immediately involving himself once the work area has been established, and necessary tools secured.

The important thing is that the papermaker respect the tools of the craft. Keep work areas clean. Lock up containers of household bleach, lye, and slaked quick-lime. It should also be noted that the use of chemicals, in many cases, can be avoided, especially when children join in the papermaking process. The extra time necessary for boiling and further breaking down the vegetable matter is small compensation for the safety of others.

MATERIALS

A list of materials and substances suitable for beating is practically endless. The selection may come from readily available papers such as butcher wrap, watercolor paper, construction paper, blotters, and rice papers. All of these break down in an electric blender with little resistance. The home or classroom should contain many disposable odds and ends of assorted papers that may be useful in beginning experiments. Printing inks will be washed out of the paper at almost every step of the pulping and bleaching process. The inks will impart color to the pulp which may or may not be desirable. A small amount of bleaching liquor (household bleach plus water) will eliminate most dyes from preprinted material being considered for recycling into new paper. However, bleaching will weaken paper fiber and should be used with discretion.

The papermaker's work table showing trays, blender, and assorted hand tools necessary for making paper. Protective plastic covering on the floor helps to keep work areas clean.

Excellent raw material is available in the form of vegetable fiber. During the fall and early winter months, garden growth is often cut back, trees are trimmed, and dead plants are set aside for the compost heap. This can be an opportune time for the papermaker to gather materials such as cornstalks, gladiolas, and iris leaves which can be blended for an assortment of papers unrivaled for artistic use. During the late summer wild plants in cpen fields such as cow parsley, nettle, esparto, straw, wheat stalks, leaves of rushes, and coarse grasses may be used individually or mixed to form a pastiche of summer growth. The scent of the finished paper, after it has been dried, is fragrant and adds another pleasure to the craft.

Surprisingly large quantities of raw vegetable material are needed to make a few sheets of 8-by-10-inch paper. Several large bags of stock should be gathered and set aside in the studio or garage. Each bag should be labeled for future identification. Many papers end up resembling one another, and unless notes are taken, useful research and time spent on material forages may be lost. By banking raw material the artist can have a supply of fibers which ordinarily would not be available during the months of the year when plant growth is scarce and the ground is snow-covered.

Ingredients used for bleaching, sizing, and dyeing should be arranged on a separate shelf where they cannot inadvertently contaminate pulp and finished sheets. They should not be accessible to younger children or within reach of persons who might not be familiar with their use.

You may wish to order a few materials in bulk. Most suppliers will ship directly on a prepaid basis. A few suppliers require sizable purchases before entertaining mail-order requests. It is best to write ahead soliciting information on ordering and payment procedures. Items such as rosin, alum, and powdered gelatin go a long way. Experiment first before deciding to order large amounts.

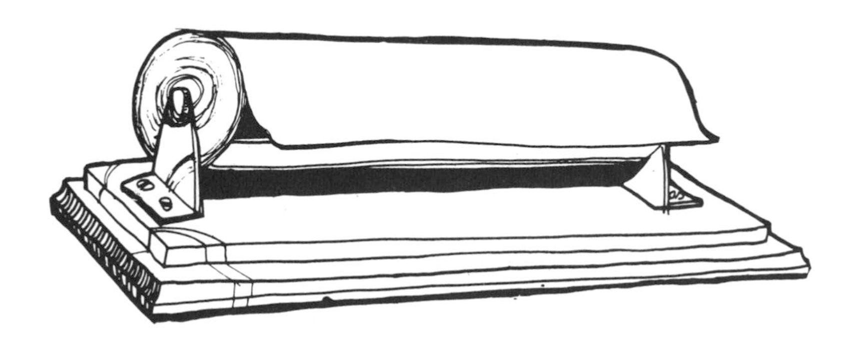

A roll of Kraft paper or butcher wrap commonly found in most classrooms where art instruction is given.

Materials

Any natural material composed of cellulose and capable of being reduced to pulplike consistency can be considered as appropriate material for experimentation, including:

Blotter paper
Paper towels
Construction paper
Paper bags
Jute
Iris leaves
Gladiola leaves
Bamboo leaves
Potatoes
Flax
Soft particle board
Mulberry
Corn stalks
Cattail
Elm leaves
Eucalyptus
Hemp
Esparto grass
Banana leaves
St. John's wort
Hay
Grape vines
Hollyhock
Bran
Moss
Printed waste
Rushes
Scotch ferns
Wheat straw
Nettles
Sawdust
Water broom
Cotton stalks
Hibiscus
Artichoke
Hornet's nest
Burdock
Cabbage stumps

Bleaching and Sizing

Alum
Gelatin, powdered
Corn starch
Lye
Plastic sprays
Rosin powder

Dyes

Procion dyes
Acrylic paints
Commercial fabric dyes
Inks

TOOLS

Small tools are necessary to prepare the stock for beating and are essential at every step of the papermaking process. About three-quarters of the tool list are items that are presently found in the home pantry and well-stocked classroom. Scissors, knives, garden clippers, and other bladed instruments must be kept sharp to function properly.

Plastic and stainless steel bowls and containers should be used whenever possible. Breakage, rust, and cost factors can be minimized with inexpensive but functional pails, measuring cups, colanders and funnels. This also serves to put more items within reach of younger artists in the classroom. At the end of a papermaking session, the cleanup of plastic and rubberized utensils is less work than having to cope with corrosive vats, tubs, and pails.

The mold and deckle are discussed under a separate heading.

Tools

Scissors
Large knife
Small knife
Garden clippers
Wood spatulas (2)
Tweezers
Funnel
Measuring jar, quart size
Colander
Bowls in three graduate sizes
Vegetable shredder
Large stewing pot, five-gallon size
Soup spoon
Measuring spoon
Plastic pails, quart to five-gallon size
Meat grinder
Mold and Deckle
Large vat or tub
Felts, pressed (from yardage stores or felt manufacturers)
Razor blades, single edge
Graduate glass cylinder
Masking tape
Grease crayon
Apron
Paper towels

Miscellaneous tools and equipment to be used in the papermaking process. All metal items should be rust preventive due to possible contamination of pulp and paper. This can be prevented with use of stainless steel or plastic items.

EQUIPMENT

An electric blender is perhaps the most essential item in the small papermaker's lab. It is used to tear, shred, and in its own way, reduce raw material to the pulp state suitable for sheet formation. If pulp is to be poured over a three-dimensional mold it must also be in a semiliquid form. Today, it is possible to purchase a good home blender with several speeds at a moderate price.

Some blenders are more desirable than others from the standpoint of the papermaker versus that of the homemaker. The blender pitcher may be glass or plastic. Plastic will scratch with the scouring process often required when pulp dries on the inner lining. The lid should fit snugly with no play and still be able to be easily removed. The electrical cord must be properly insulated. Speeds should vary from low (grind, stir, mix) to high (blend, liquefy). When the first portion of semidry material is introduced into the blender, a significant amount of power is essential to break down the substance to a finer consistency. If the blending process is not performed gradually, a strain is put on the motor which can be beyond its capability, causing it to burn out. Use small amounts in the beginning.

The electric handmixer is a more convenient version of the traditional egg beater. Both serve the same purpose. The pulp in the vat must be agitated, sufficiently breaking up the clumps of fiber that often form when pulp is liquid and float or settle at the bottom of the basin.

The hot plate may be of the gas or electric type. The gas burner requires less heating time; however, a gas outlet may not be near the work area for easy installation. In this instance the electric hot plate will perform equally well.

If it is undesirable to use an electric or gas hot plate or stove, an electric cooker may be substituted. The cooker, or electric crock pot as they are sometimes called, simply boils the raw material. Water is added to pulp, then placed into the cooker and left to simmer at low temperature for a specified period of time. The cooker is entirely self contained while the temperature is closely regulated. The drawback is its limited capacity for boiling larger amounts of raw material.

Electric juicers work on the principle of extracting juice from pulp by way of centrifugal force. Ordinarily the fruit and vegetable juices are saved and the remaining pulp discarded. However, the pulp here can be squeezed together into balls and stored in small plastic bags. At a later time the wadded substance is placed in the cauldron, boiled, and further broken down. The selection of vegetable material is almost unlimited. Artichoke leaves, celery, carrots, kale, and chard can be processed through the juicer-extractor. The yield is small but colorful, and interesting in texture. This is a good exercise in developing paper on an experimental basis. If the results are to the artist's liking, then he may proceed to make larger quantities following other steps described in the book.

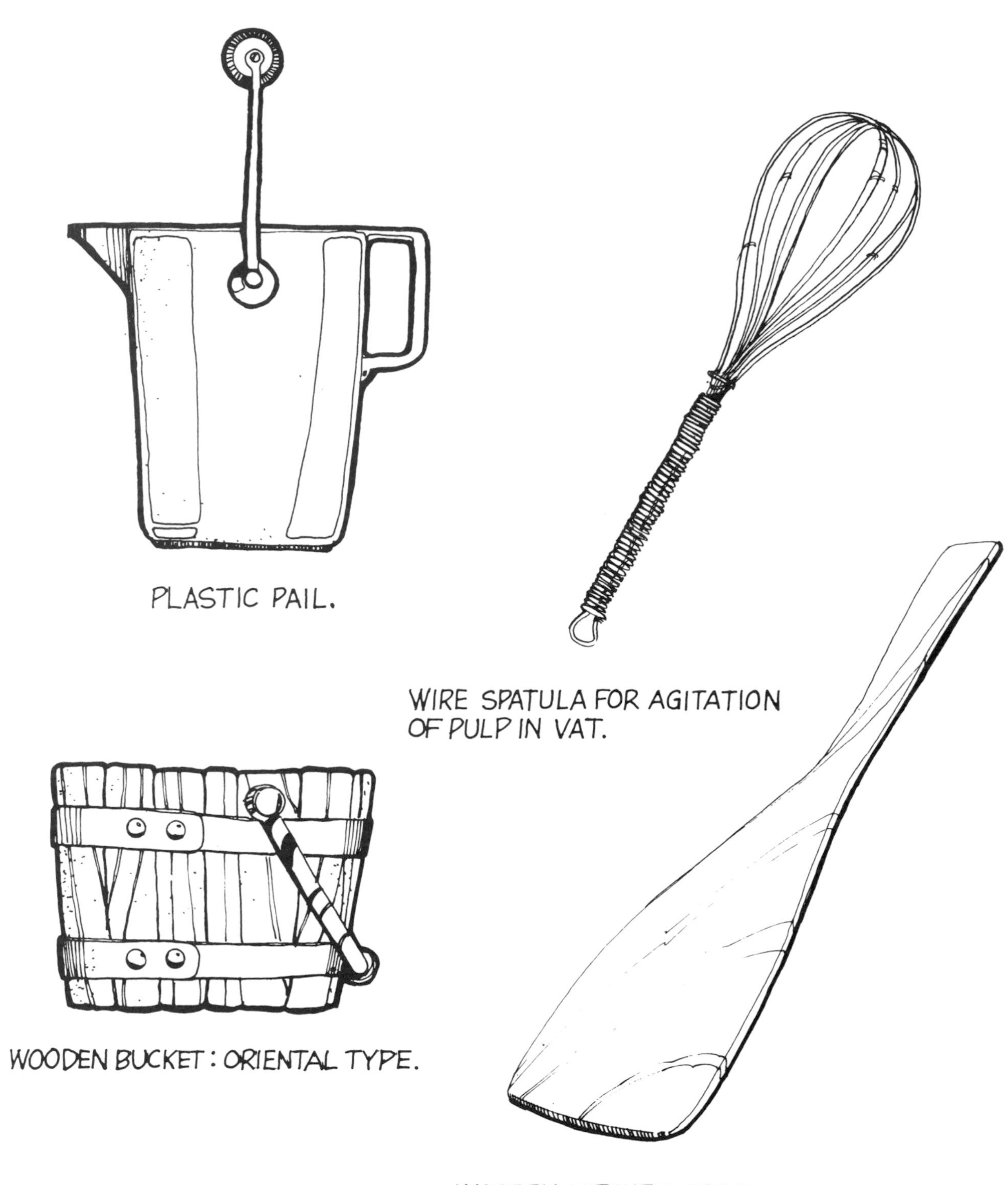

Various household items used in the papermaking process.

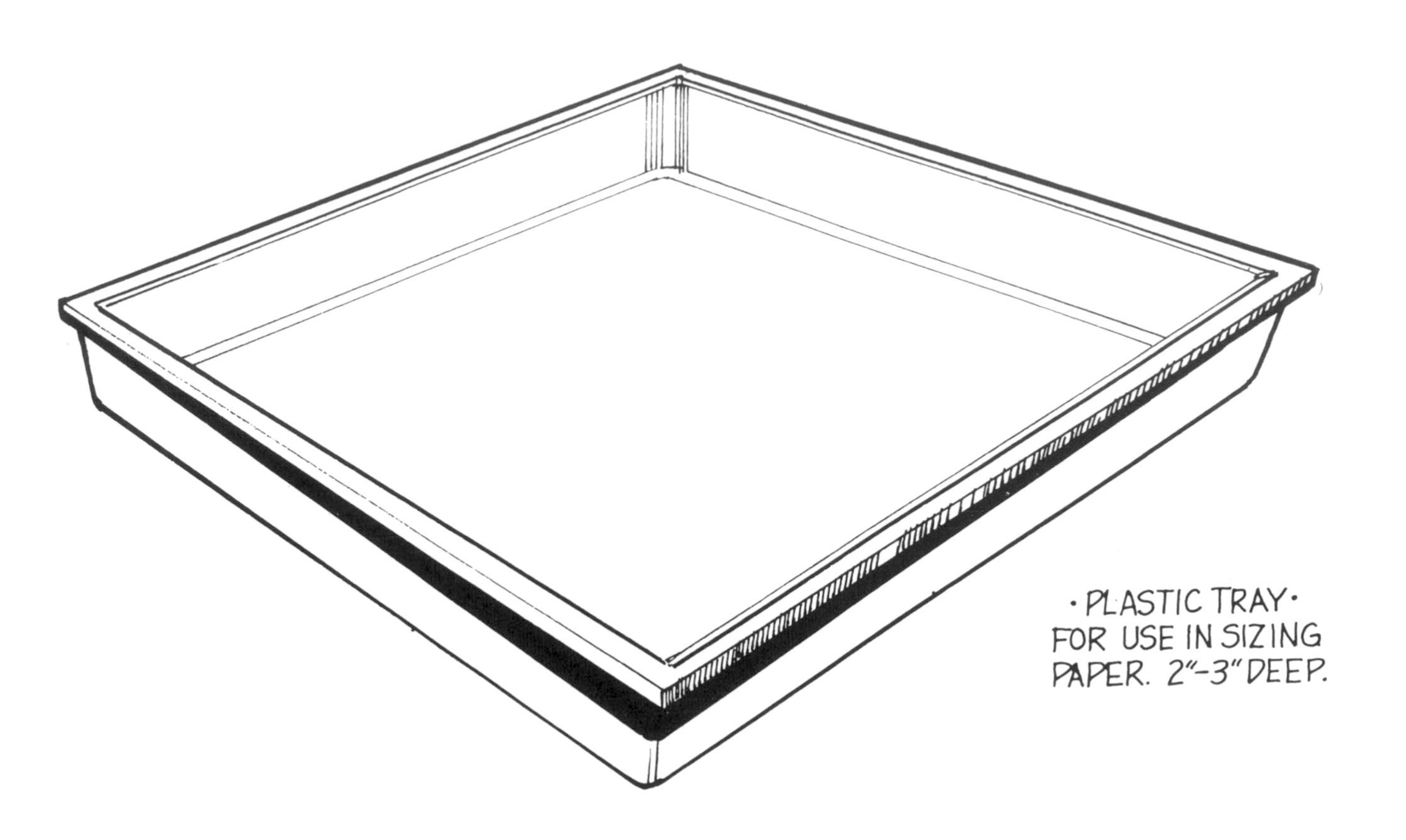

· PLASTIC TRAY ·
FOR USE IN SIZING
PAPER. 2"-3" DEEP.

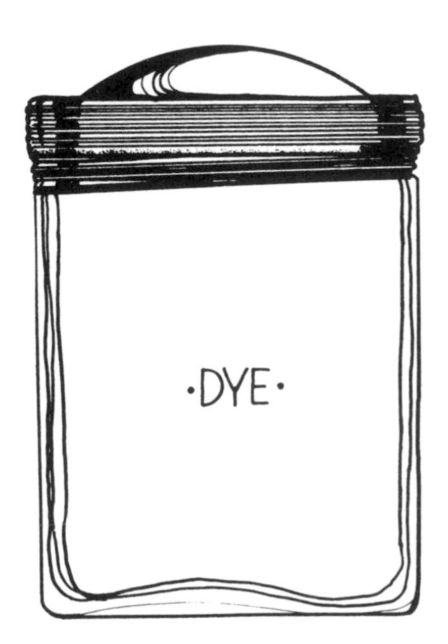

GLASS JAR FOR STORAGE
OF DYES & PIGMENTS.

An inexpensive meat grinder with assorted blades. The grinder is capable of reducing small amounts of some grasses and vegetable fiber to a state suitable for boiling. After boiling the pulp is made.

Equipment
Iron, electric
Blender, household type with several speeds
Electric mixer
Hand mixer or beater
Hot plate or burner
Gas burner, one or two plates
Press, wooden or book press
Meat grinder, hand-operated

Optional
Juicer which extracts juice from pulp with centrifugal force upon maceration of fiber.
Electric cooker or deep fryer, sometimes referred to as electric crock pot.
Hydraulic press. This device is capable of exerting extreme pressure on the post for removing excess moisture.
Print dryer.

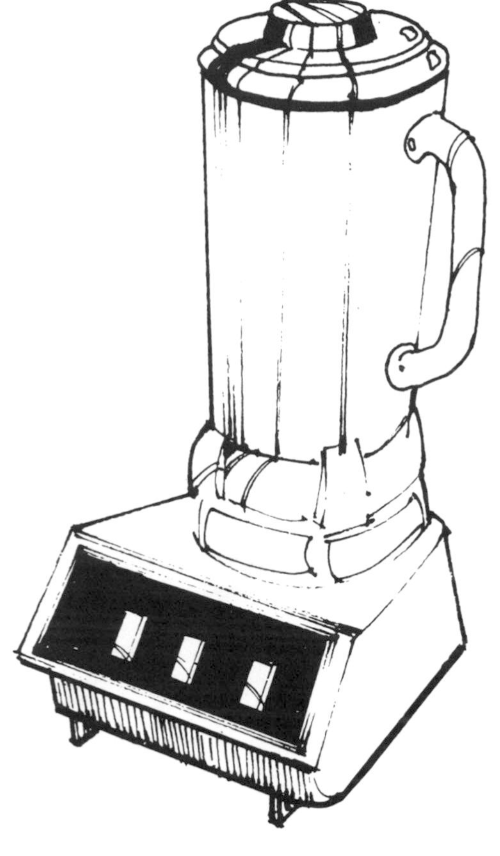

An electric blender equipped with several speeds.

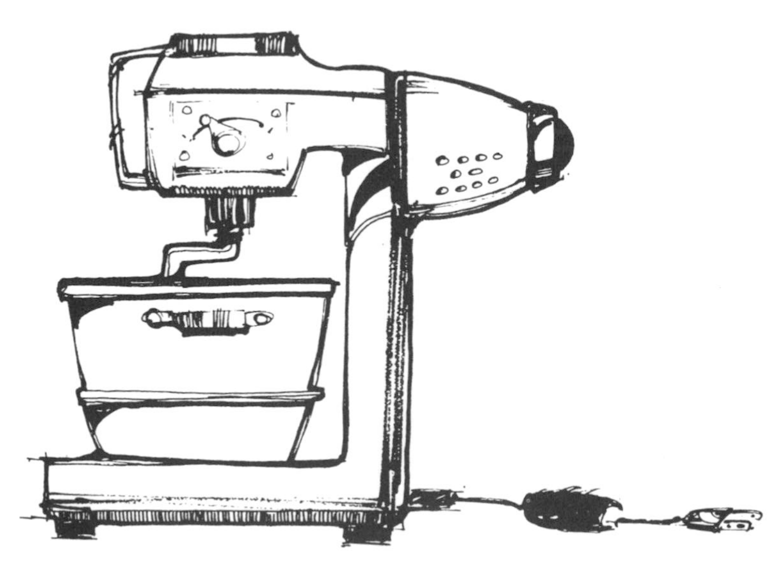

Mixer used to macerate simple grass fiber.

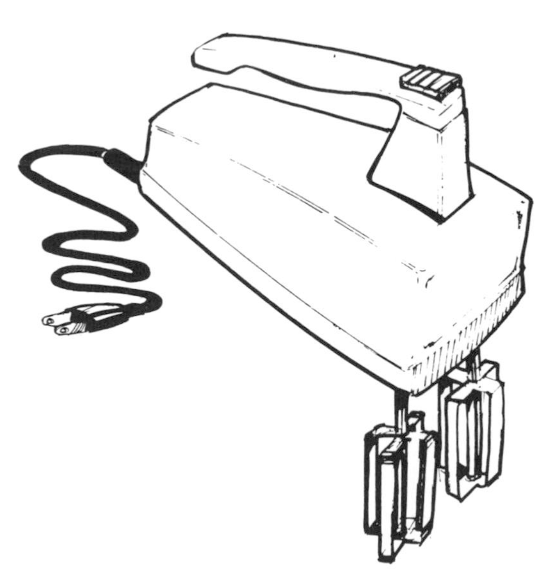

Electric beater used to agitate the pulp in the vat just prior to forming.

THE MOLD AND DECKLE

The mold is a most important piece of equipment for the papermaker's use. It is the tool that actually makes the sheets of paper. Usually rectangular in shape, the mold consists of a screen with a wooden edge resembling a small version of a modern window screen. On top of this is usually placed a second, separate wooden frame called the deckle. The deckle is used when the pulp is to be confined within the surface area of the screen. It aids in determining the size of the individual sheet.

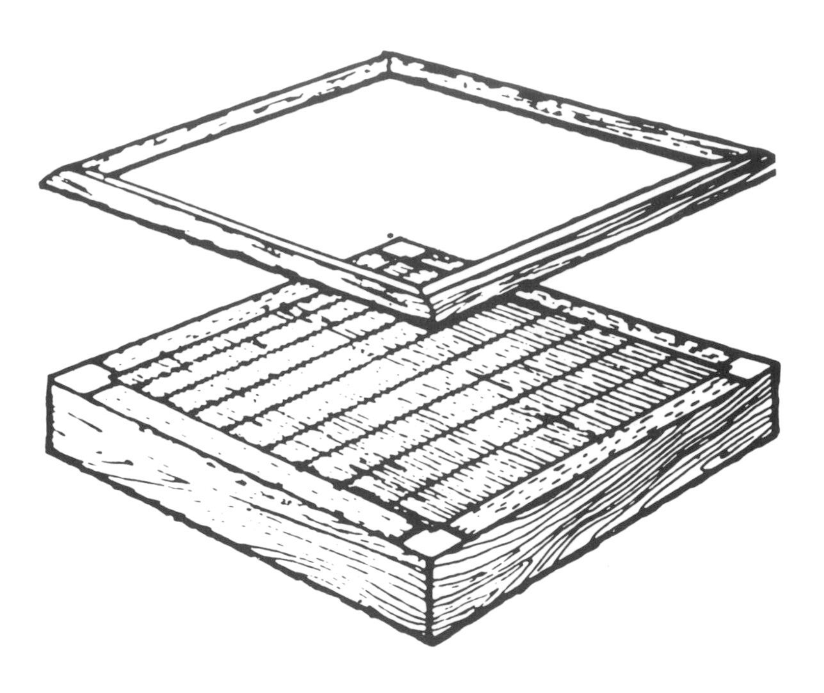

A mold and deckle. The deckle is fitted over the mold to prevent excess pulp from spilling over the sides of the mold as it is drawn from the vat.

A handmade mold and fitted deckle. The wood is mahogany.

The mold and its screenlike material are best made from mahogany and brass or bronze wire. In the Orient early molds were made using cloth and eventually fine strands of bamboo. However, lacking this most precious material, Europeans designed the "rigid" wire mold consisting of the two parts mentioned above. The deckle must fit exactly and be raised off the surface of the screen to properly contain the thin layer of fibrous pulp preventing it from running over. The thickness of the deckle is determined by the substance of the sheet to be formed. The term "deckle-edge" refers to the distinctive irregular edges of a sheet of handmade paper. No matter how tightly the deckle fits, a small amount of the thinner pulp will creep under the frame. When the sheet of paper is dry, the serrated edge is seen. Normally, this edge is trimmed with a blade or paper cutter. This also takes away much of the handmade paper look. Bookbinders and artists today often retain this deckle feature, a sure sign of quality and fine craftsmanship in handmade paper.

Almost all sheets of paper are made with the "wove" or "laid" finish. By holding a sheet of paper to a strong light you will almost immediately see the laid finish which is characterized by faint lines caused by the manner in which the sieve of the screen was fashioned. The laid screen was actually sewn into the mold frame by a series of intricate and painstaking techniques usually best accomplished by the mold specialist. Wove paper, however, is perhaps best made when the mesh of the

A thin piece of handmade paper showing the deckled edge characteristic of hand-molded paper. This particular sheet was made out of dried bamboo leaves.

mold is made using finely woven cloth or wire. This type of mold imparts to the paper sheet a mottled effect which evinces no lines or other marks characteristic of the mold. The use of woven cloth or other material is actually as old as papermaking itself. The Chinese used a woven fabric on their first molds, but it was not until the eighteenth century that the famous printer and type founder, John Baskerville, demanded sheets of paper with no visible lines and a smooth surface. At present, handmade papers are fashioned using both types of finishes, each providing the papermaker with its own particular look and surface.

A section of paper from Fabriano revealing laid lines and watermark.

Bottom view of a simple mold which may have fabric or wire stretched over the top of the frame to support the pulp.

The wove mold is the easiest to make, often being fashioned out of handy and interesting fabrics found in a yard goods store. Here the material is taughtly stretched and stapled onto the wood frame using brass staples (steel staples rust). You may wish to experiment with ready made plastic screen door mesh available in hardware stores. This material, by virtue of its plastic composition, will not rust and is inexpensive. Like the fabric, plastic mesh should be reasonably open in its weave. Material that is too fine makes paper that is very thin and irregular in thickness. The center of the mold holds more pulp making this area of the paper too thick while the peripheral areas are too thin. When selecting cloth or plastic screen, look for material that is approximately 30 mesh, which means 30 strands of material to the inch. This will give your paper an even, consistent appearance and should be the most trouble-free size to use.

Ideally, the best molds with the longest life are those that are fashioned out of woven brass. This wire may also be 30 mesh. Other mesh sizes will work well too (the author has and continues to use screens with 60, 80, and even 100 mesh for papers of great delicacy). Here, as in other instances, you should experiment to find what you like best. Woven brass wire is relatively expensive, and not always easy to find. A 12-inch square may cost as much as $4.00 to $6.00. When a full roll of five feet is purchased the price per square foot usually drops. This length will enable you to make three to five small molds and, for the serious papermaker, is a good investment.

In contrast to the high cost of brass wire, it should be emphasized that good, quality paper can be made with even the most crude and primitive of materials provided sound and proper techniques of construction are followed.

Once the decision on material is made it will be necessary to build a frame to support it. The frame is made of wood which will withstand the alternating soaking and drying that occurs when the mold is in use. Traditional molds are made of mahogany or teak. The unique properties of these woods make them ideal for use in water without benefit of varnishes or sealers. Other woods such as pine, fir, ash, or walnut will stand up to repeated use and, for the hobbyist, are perhaps more desirable because they are cheaper and more readily available. Whatever material is used, it must be free of knots and warpage. A mold that is true will give years of reliable service. The following are two simple methods of constructing practical molds and deckles.

Method 1

A rectangular wooden frame with a piece of fabric nailed or stapled across it makes a good mold of useful shape. Without the aid of the deckle, the paper is allowed to dry on the surface. Serveral molds make several sheets of paper during one session. On a warm day the paper is placed outside to dry and will do so after four or five hours of direct exposure to the sunlight.

Wood for the frame should be 1-by-2-inch pine strips clear of any knots or irregularities. Two 8-inch strips and two 13-inch strips are joined together to make the frame. The corners are mitered at 45-degree angles and screwed with ¾-inch flat-head wooden screws. Flat L-shaped metal braces should be applied to the underside for greater strength and rigidity. These must be brass, as should all hardware used throughout the mold-making process. "Foxing" can occur anywhere and invariably shows up in the handmade sheets.

Once the frame has been secured, material is stretched taughtly over the frame's edge, fastened with brass nails, or staples, and then excess material is trimmed off. This simple mold may be used with a good degree of success. If extended life is to be obtained from the mold, copper or brass strips should be placed on top of the mold's edge to further minimize fraying. Small strips of ¼-inch dowel may be placed on the underside of the frame to gently support the screen material. When the mold is lifted from the vat, a strong suction draws the fibrous material towards the middle of the mold where the least resistance occurs. The result is paper of uneven surface and thickness. In your search for suitable material, consider using fabric such as printmaker's tarlatan, coarse cheesecloth, burlap and other open, yet durable, material.

Detail showing position of an L-shaped flat brace. In a nonmitered corner the brace helps to secure both pieces of the mold frame.

Method 2

The basic design of a more professional mold of the wove type may interest the more serious papermaker. The rectangular frame of the mold is made of mahogany or oak to the required size of the sheet. On top of the frame, brass or bronze woven wire is placed and securely fastened with small brass brads approximately ½ inch in length and spaced every ¼ inch. On the underside of the frame, wedge-shaped ribs are spaced 1¼ inches apart parallel to the shortest sides of the frame. The wedge shape makes it easier to draw the mold from the vat then it would be if it was flat. The wire mesh is secured by copper stripping around the frame and additional brass brads. Flat L-shaped braces are then fastened to the bottom four corners. If a watermark is desired, make it of very fine brass wire, sew the wire shape to the mesh with very fine wire thread, and trim waste ends. Wire which is too thick will weaken the structure of the paper and is undesirable in cases where the paper is to be printed.

An example of paper formed with the use of a round sieve with 80-mesh screen. The pulp is poured into the sieve and allowed to dry in this position.

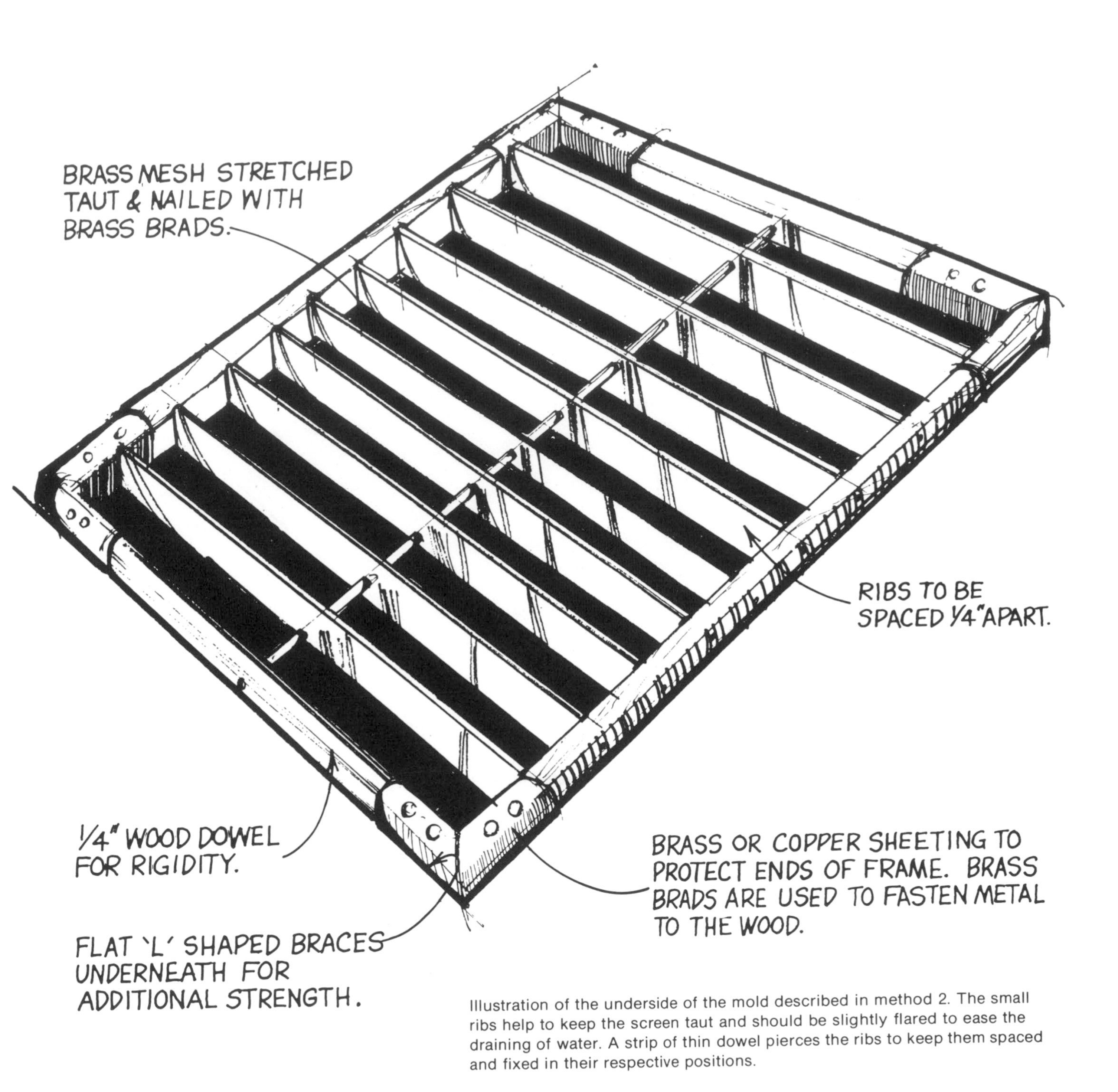

Illustration of the underside of the mold described in method 2. The small ribs help to keep the screen taut and should be slightly flared to ease the draining of water. A strip of thin dowel pierces the ribs to keep them spaced and fixed in their respective positions.

The Deckle

The deckle is the removable wooden frame which fits the mold exactly and creates a raised edge. A suitable deckle may be constructed of picture frame molding of approximately 1 inch width with a recessed underside. The molding is cut with a miter box and fastened together so that it can easily rest upon the mold. All four corners are capped with copper stripping and secured with small brass brads. If glues are to be used in construction of either the mold or deckle they should be insoluble in water. Animal-hide glues and synthetic glues created with water repellant properties are usually available in hardware stores.

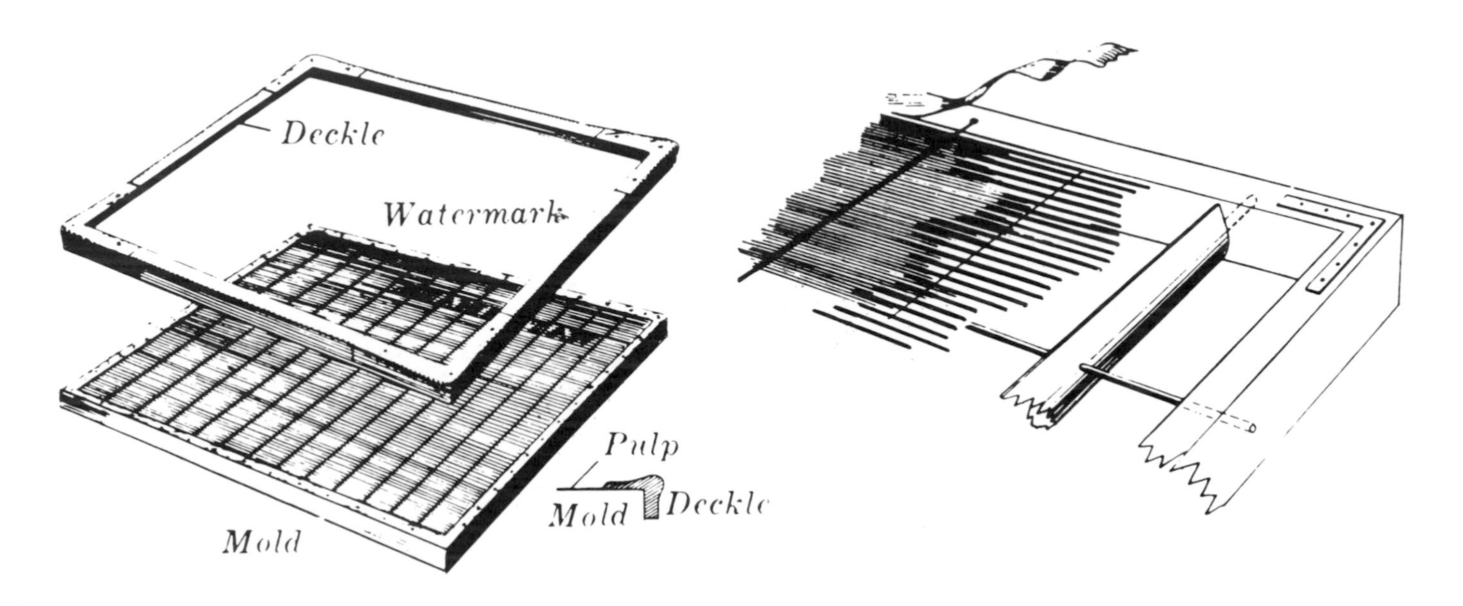

Two views of a more professional mold and deckle typical of the ones fashioned in Europe by master craftsmen. Only the finest materials are used to give extended life to the product. (Quentin Fiore, *Industrial Design*, 1958)

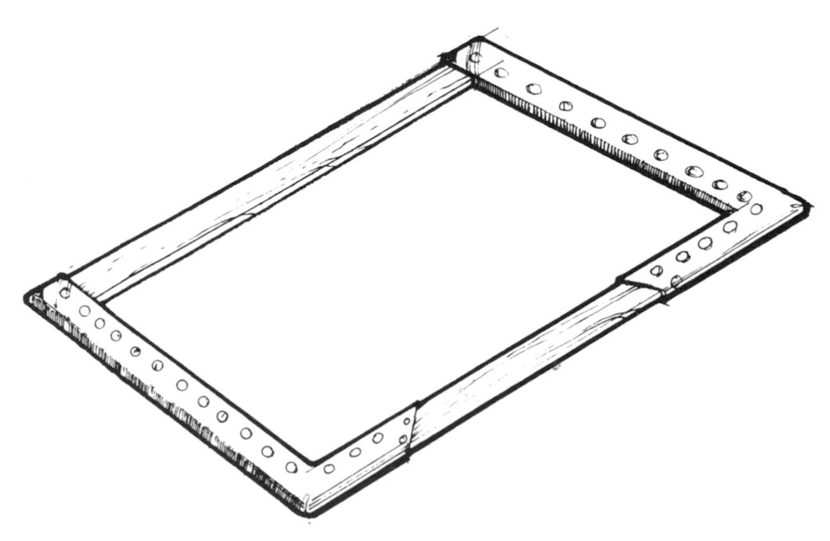

A simple deckle fashioned of teak and copper strips. Brass brads are used to fasten the copper to the wood.

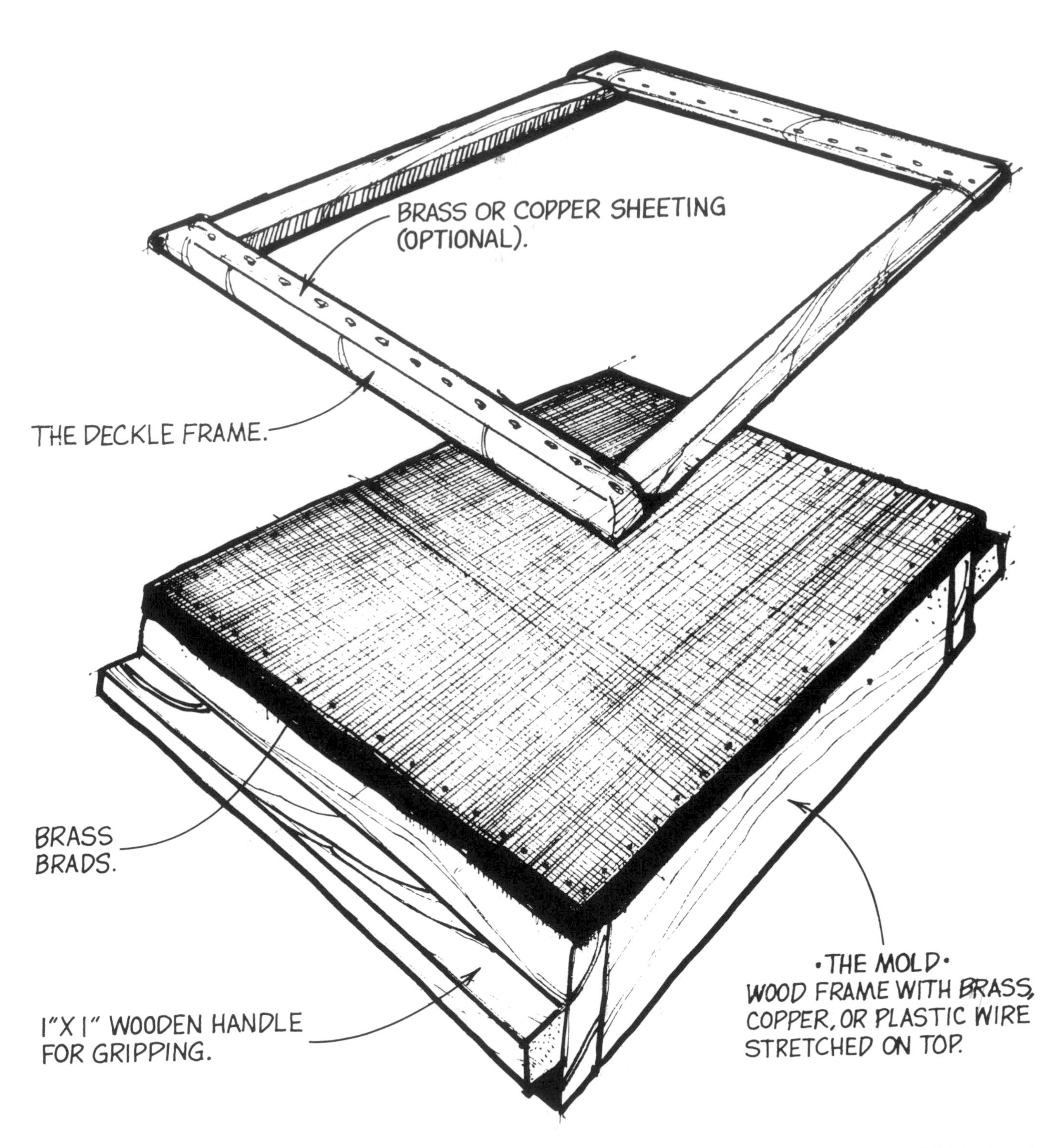

The fabric or wire wove screen fitted over the frame. On top of this is placed the deckle.

HOLLANDER BEATER

The serious, experienced papermaker may wish to add more costly items to extend the range, type, and amount of paper he can make. The most expensive item within the papermaker's studio is the beater. The beater (Hollander type) is a machine that came into extensive use in the 1600s in Holland and is so named for its place of origin. It is rather simple in construction and function. A cylinder with metal blades is set into an elliptical tub, usually of copper or stainless steel, and grinds the cotton rags, which have been cut into small pieces approximately one-inch square, against an adjustable bed-plate. The revolving motion of the roller is regulated by a small,

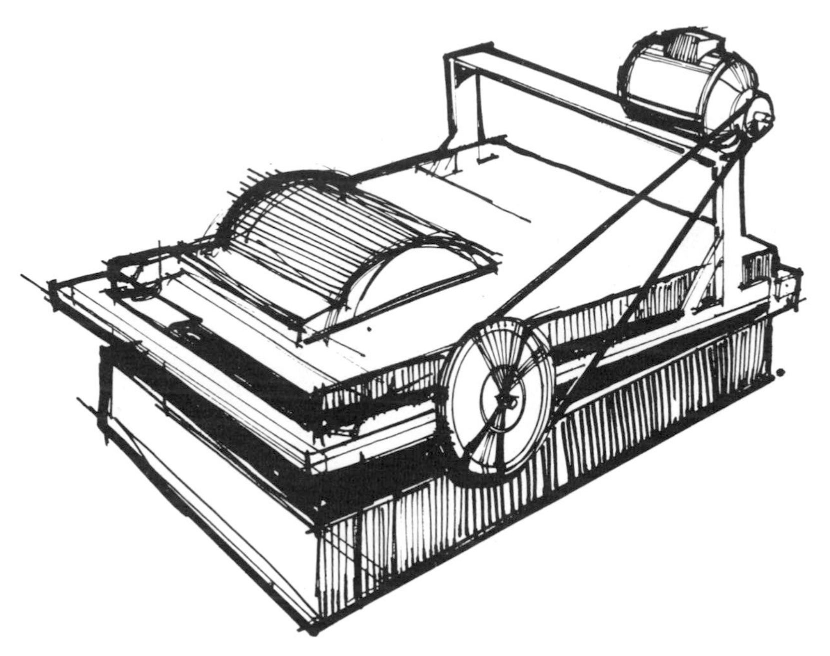

A simplified version of the Hollander beater showing motor and pulley with vat enclosed by copper lid.

Side view of beater.

low horsepower motor mounted on one side. The roller keeps the pulp circulating continuously around the tub, passing under the blades until the desired degree of maceration is attained. This machine, both in principle and function, is the workhorse of the papermaking industry today.

Needless to say, if purchased, this piece of equipment is prohibitive in cost and will no doubt be outside the means of most amateur enthusiasts.

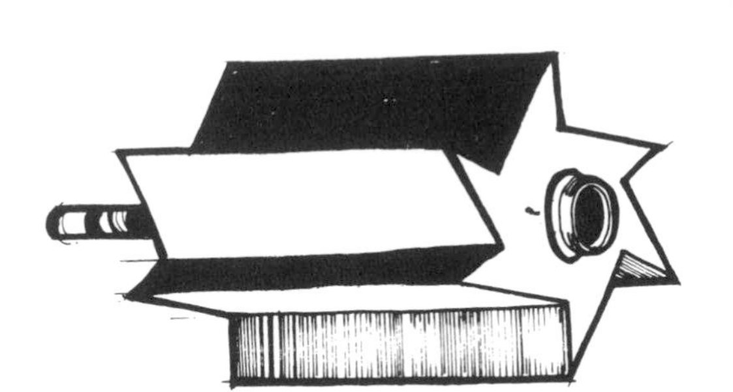

The bladed cylinder used to macerate the raw fiber.

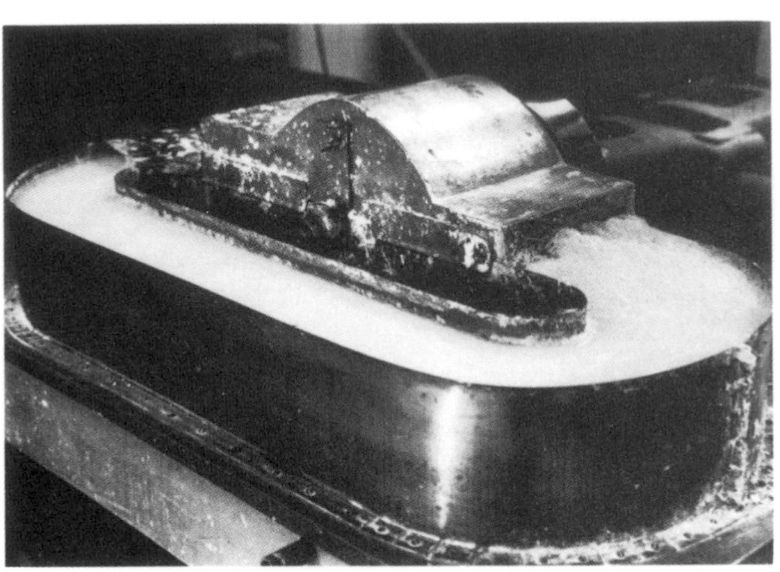

View of the beater with maceration of pulp underway. The pulp moves around the trough in a circular motion as the material passes under the blades which grind or "beat" the fiber

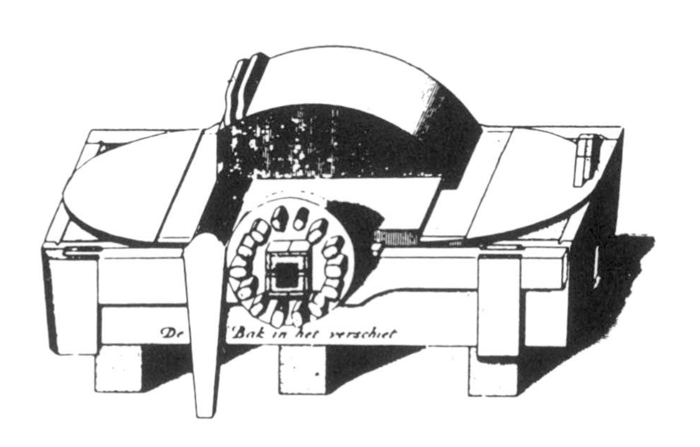

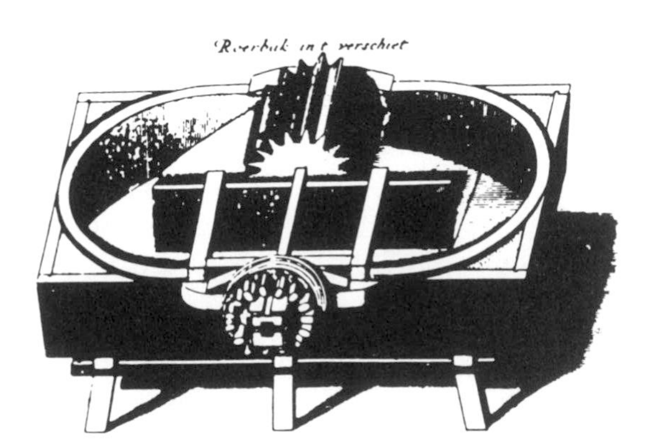

Early illustration of the Hollander beater showing covered and uncovered view.

4. How to Make Paper

Millions of years ago, there evolved on the earth's surface a lowly insect known as *Hymenopterous* of the family *Vespidae*, more commonly known as the paper wasp. This little creature, innocent enough in appearance, keeping most people at bay, is capable of creating a most extraordinary type of paper. This miniature papermaker, without benefit of vats, molds, and beaters, nibbles away at a minute piece of raw wood and, through its own digestive process, is able to transform raw material into pastelike substance, resembling paper, and uses it to create its nest habitat. Man, the papermaker, is capable of creating an almost identical product but unfortunately must rely on the necessary accoutrements of the trade: beater, mold, deckle, vat, couching felts, and a press. It is with the beater and its function that the papermaking technique we know and use today begins.

There is little romance to the beating process. Once the raw materials have been gathered they must be broken down into small fibrils capable of interlacing themselves when the mold and deckle are lifted from the vat. There are several labor-saving devices for breaking down fibers, which have been in use for centuries. The simplest still is the hammer and hollowed-out log or stone. The vegetable material is placed in the concavity of the object and then a hardwood mallet is employed to beat or macerate the material until it resembles a pulpy mass. It is then treated to a vigorous boiling with a combination of water and caustic solution such as lye or slaked lime. The pulp is thoroughly washed several times, and often reduced to an even finer state with subsequent beating or shredding.

Once the desired coarseness of pulp is obtained, the material is distributed within the vat. The mold and deckle lifts the pulp from the vat leaving a thin stratum of shiny fibers on the surface. The damp sheet is couched onto felts and dried. A closer examination of each step in the above process will help the beginning papermaker to successfully prepare the stuff with little difficulty.

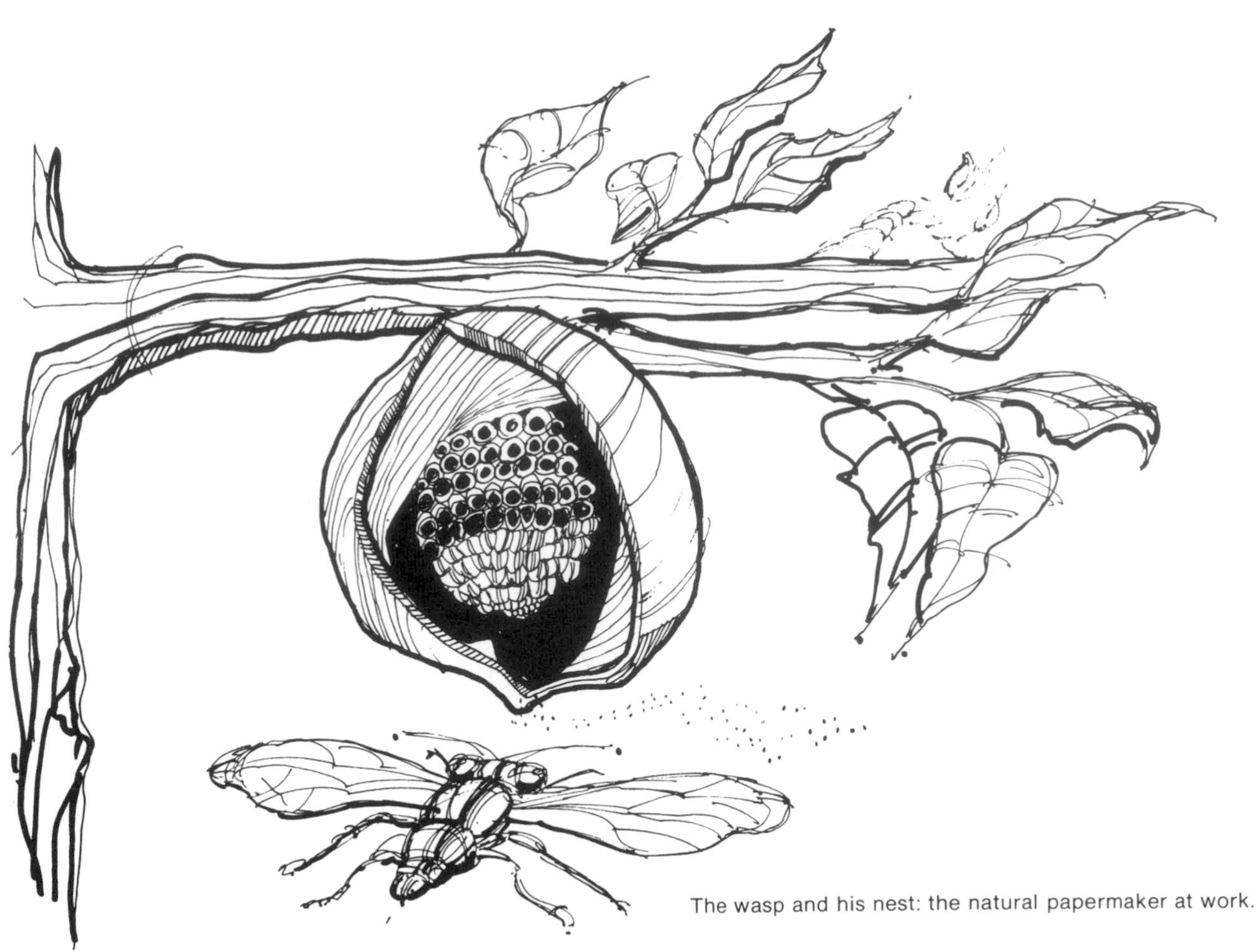

The wasp and his nest: the natural papermaker at work.

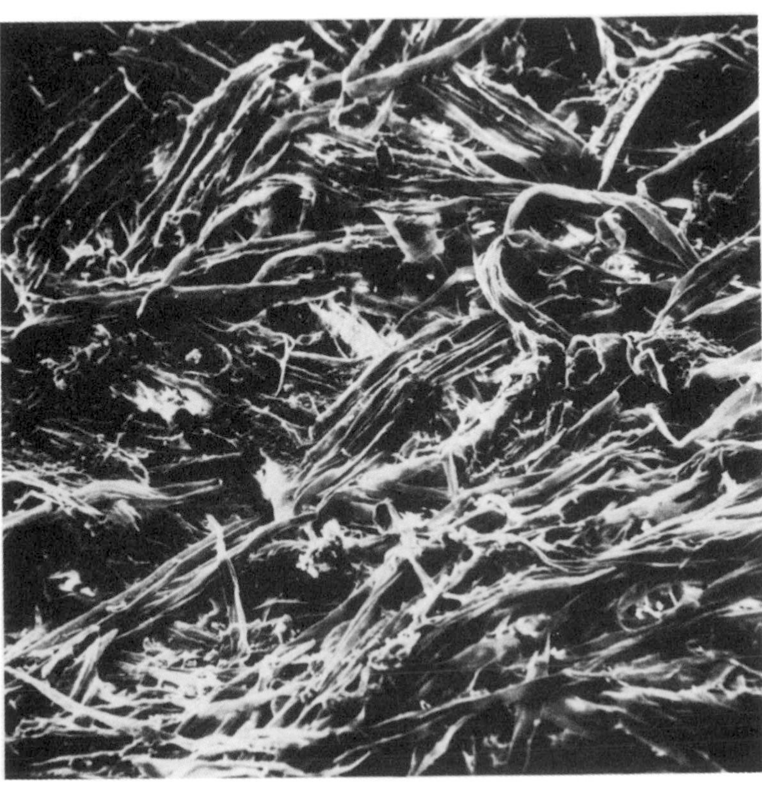

Scanning Electron Micrograph of the wasp's nest magnified 320 times. Notice the similarity between this material and the photographs at the front of the book. (Courtesy of the Industrial Laboratory, Eastman Kodak Company)

SCISSORING THE RAW MATERIAL

It is the skeleton of the plant that is used for paper. The fleshy parts must be removed by boiling. First the raw plants are cut into ½-inch strips with scissors, shears, or clippers. This may also be accomplished with a small gasoline operated garden shredder so often employed in making compost heaps.

The raw material (cornstalks) before it has been cut. The stalks are thoroughly dried and left to become brittle before being used.

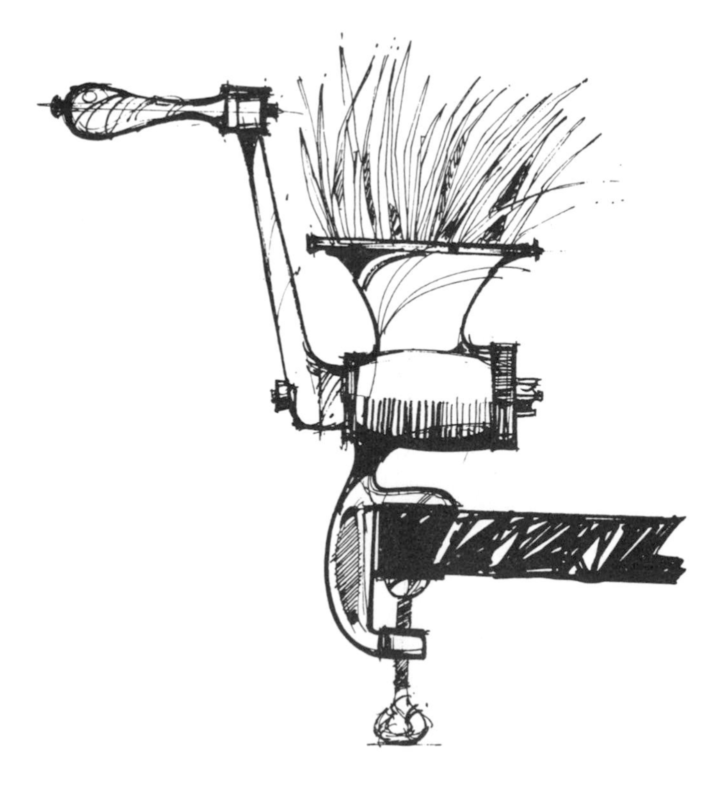

Stalks, grasses, straw, and dried flower leaves may be ground in small quantities by use of the meat grinder.

This device quickly shortens the time necessary to do a thorough job. Hand scissoring is best on delicate material such as iris, cornstalks, and gladiolas. An ordinary hand-operated household meat grinder may also be used for light shredding. Be sure to feed the stock slowly into the grinder and in small amounts.

More difficult, tougher material should be cut with long hedge clippers. In extreme cases, a woodsman's hatchet can make short work of resistant materials.

Small, delicate stalks may be shredded by a common vegetable shredder found in the kitchen.

Material being scissored into small pieces. The smaller pieces will then be placed into the pot for boiling.

BOILING THE PULP

Once the fibers have been shortened, they are placed into a four-to-five-gallon baked enamel cooking pot. Avoid any material or tool that may rust, spoiling paper with pollutants that will cause "foxing" (small, brown spots often found in papers where corrosive elements have inadvertently come into contact with pulp or paper). Fill the pot with the dry cuttings approximately halfway. Add an equal amount of water. Be careful not to raise the volume of the contents too close to the top of the container. If you decide to use a caustic solution to aid in the reduction of the fleshy parts, the

A five-gallon porcelain-covered pot used to hold the raw material as it is boiled.

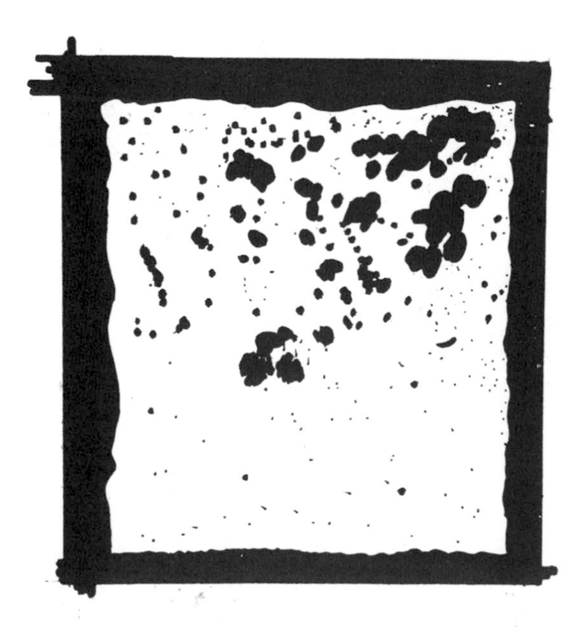

Illustration of foxing. This occurs when metal objects inadvertently come into contact with pulp or paper. The color is oxidized and resembles a brownish rust stain.

amount in any one boiling should not exceed three to four tablespoons. The chemical most readily available is lye or slaked lime. These two items are corrosive irritants which destroy tissue. The least amount necessary to perform the job should be used. Be sure to measure out the chemical and place it into the bottom of the pot before any material, water, or fiber, is to be added. If lye is added to the warm water, it can set off an immediate chemical reaction causing noxious vapors to fill the air. Read the manufacturer's instructions on the label before proceeding.

Placing the contents into the porcelain and metal pot.

Pour the water into the pot and stir the contents.

It is not always necessary to use a caustic ingredient. A longer boiling will render the material useful, devoid of fleshy parts, if this approach is preferred. Simply keep an eye on the boiling plant, occasionally lifting out a small amount, letting it cool, then squeezing it to see if it has a slippery, slightly mushy feel. Experience will allow you to determine when the first boiling has accomplished the necessary results. Throughout this procedure wear rubber gloves to protect hands and arms from irritants.

Two to three hours of boiling on a simmering flame or low temperature hotplate will sufficiently break down the material. When this time has elapsed, turn off the heat, let the pot stand and cool. Caution should be exercised when transferring the pot and contents to the sink or laundry tub. The sheer weight of it can be easily misjudged and spilling is all too likely. It may be more convenient to remove the pot to a lawn or remote part of the yard for the necessary washing.

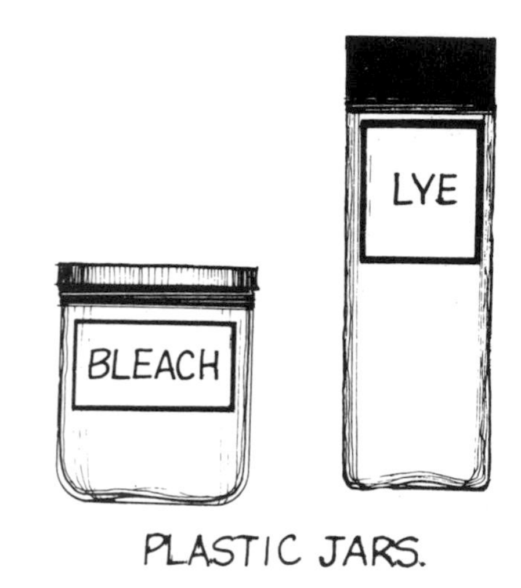

Containers identified as lye and bleach. These ingredients are used to lighten the color of the pulp.

Stir the pulp for better absorption of caustic solutions.

WASHING THE PULP

Washing the pulp aids in ridding the useful fibers of irritants and nonfibrous debris. Allow water to play gently down onto the material for no less than fifteen minutes until the water is clear and the natural color of the fiber reveals itself. This indicates that brackish elements have been eliminated. Pour out the remaining water. Notice the vast reduction in bulk compared with the initial amount of material. What now remains is the skeletal fiber of the plant, practically all of it useful.

It is recommended that the above boiling procedure be repeated once again, however, shorten the boiling period to approximately thirty or forty minutes. This usually conquers the most stubborn material. Wash once more, then place the pulp into a glass, clay, or plastic storage jar, and seal with an appropriate lid until it is time to mix it into the vat. If left to rest on a shelf for a few days to a few weeks the contents will be further broken down by decomposition. This may, at times, be desirable. Pulp is wonderfully easy to work with when in this state, but attached to this condition is often a rather pronounced odor reminiscent of pond stagnation. If this is found to be unpleasant, add one teaspoon of household bleach to every two quarts of pulp and stir. This will eliminate much of the odor.

The bleach liquor may also lighten the color of the pulp but is injurious to the fibers by structurally weakening them. If you wish to purposefully lighten the natural color of the plant, experiment for best results. In any event, it is recommended that you not exceed four tablespoons of bleach per two quarts of dense pulp. The fiber will usually lighten within a few hours. Before using the pulp, thoroughly wash it and squeeze it with your hands to remove all traces of the bleach liquor.

BEATING THE PULP

The plant material is now ready to be reduced to the consistency necessary to form paper. It has often been said that the beating of the stuff is where the paper is made. Japanese papermakers use a variety of mallets to pound the pulp, greatly shortening the fibrils. This task is usually performed by village girls, occasionally extending their labor into the early hours of the morning in order to provide the vatman with enough raw material to make a day's post of paper. This monotonous but necessary drudgery is often accompanied by a poetic ballad sung by the Japanese paper beaters: "At Simo-gyo the winter seems to have come with the dreary sound of pulp beating. . . ." Throughout Japan, small villages may be found which are entirely dedicated to the craft of papermaking. It is an unusual sight to encounter an out-of-the-way village where paper of all sizes may be seen clinging to the sides of huts, houses, enbankments, and specially constructed drying racks. The paper, on occasion, is left on the mold to dry in the long day's sunlight, only to be removed at dusk and stored for the next early light. The Japanese choice of ingredients for their pulp includes plants such as kozo, gampi, and mitsumata. These plants yield the lustrous and silken beauty that has elevated the craft of papermaking in Japan to a place of high honor in the Oriental arts.

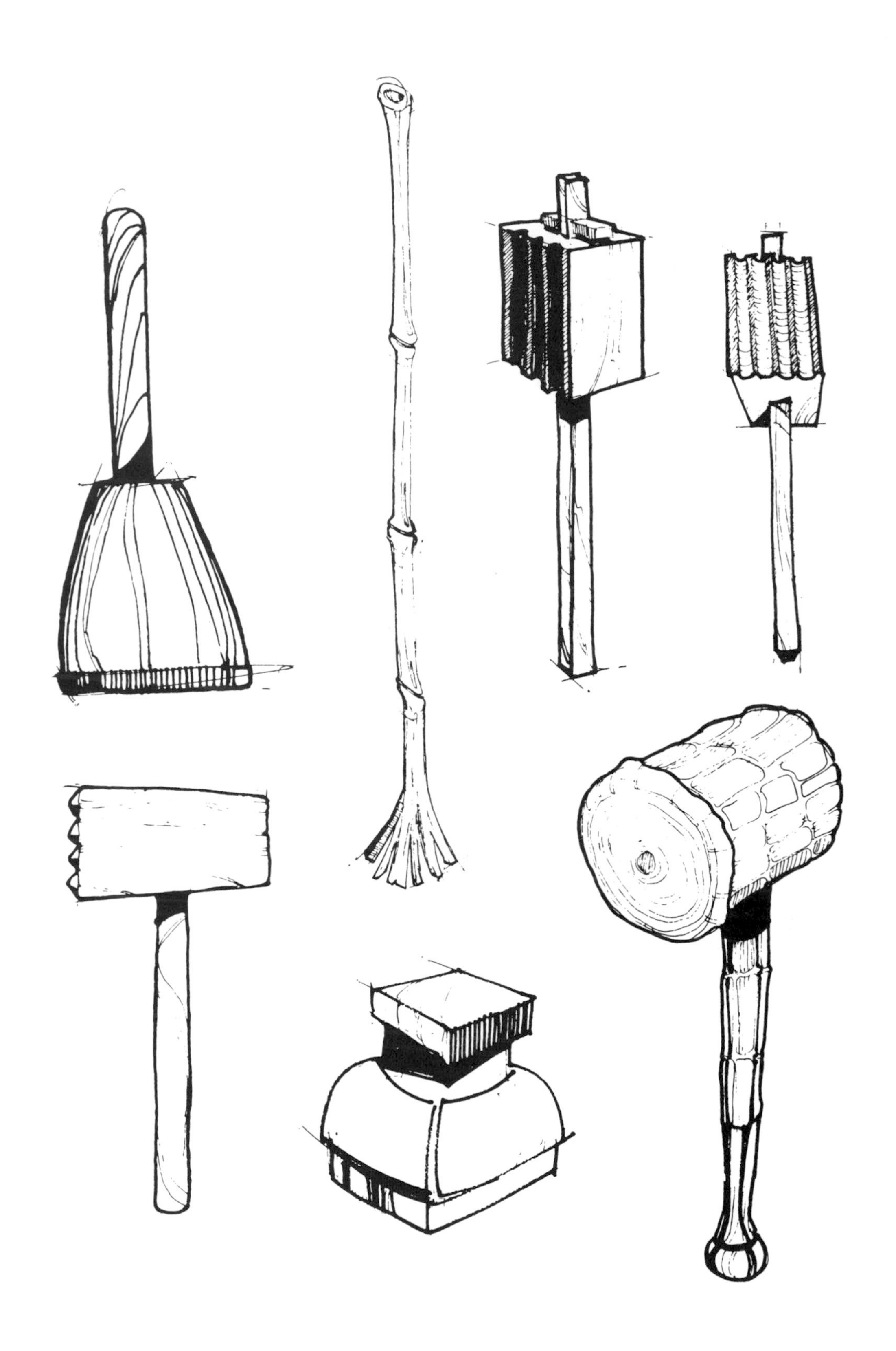

Tools used for beating the fiber to reduce the material to the pulp state. Hand beaters are usually made of wood or stone. The fleshy part of the plant is separated from the skeletal support by the rhythmic beating motion of the paper beater.

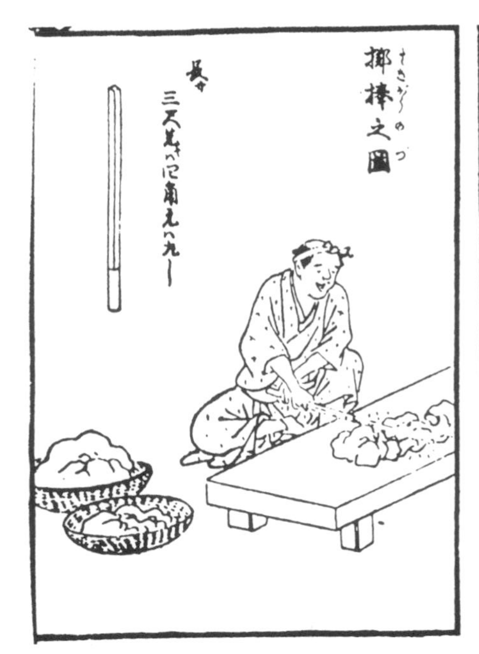

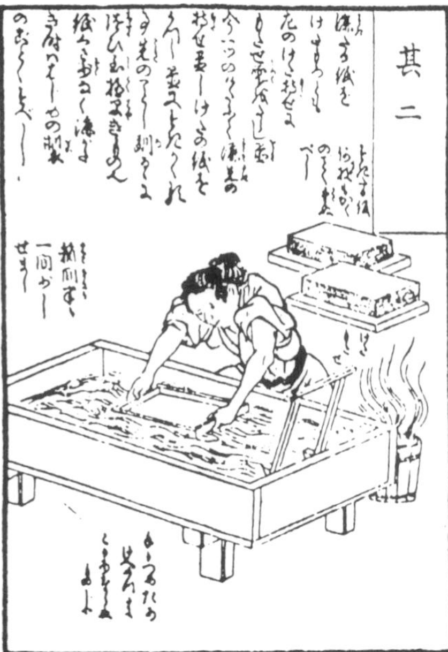

Paper as made by the Japanese. Left: The raw fiber is beaten with a three foot long stick of bamboo. Middle: Traditionally, hanshi paper is made by women because of the ease in forming the light fiber. The steaming bucket is used to warm hands. Right: Drying of the paper. Five sheets are placed on a six-foot board called the bed. (*Kamisuki Chōhōki Kunisaki Jihei*, 1798. Reprinted by the University of California, Berkeley, 1948)

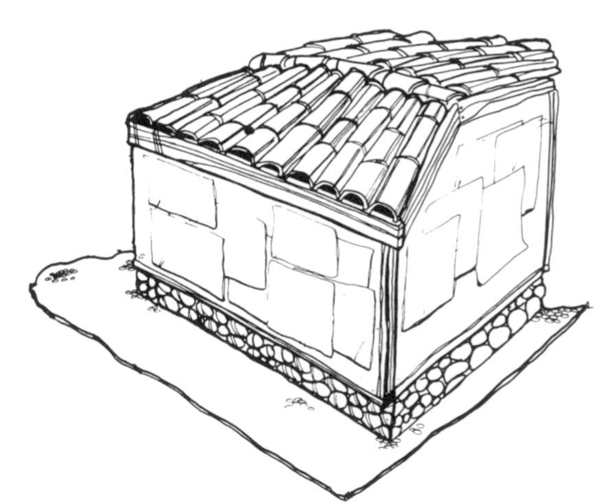

At the end of a papermaking session, the Japanese place the newly formed sheets on the sides of houses to dry. When dry, the sheets are gently peeled off and tied in bundles ready for loading.

Four of the common plants used in the manufacture of Japanese papers. Each is used for its own particular quality and finish. Left to right: mitsumata, kozo, gampi, and tororo-aoi

In the Pacific Islands, beaten bark material, commonly known as "tapa" cloth is a substance that may be given the appelation "paper." Ethnologists may disagree with this appelation by fixing the definition of true paper as having to be formed from small fibrils, thereby disqualifying a continuous flat substance such as tapa cloth as being paper in the accepted sense. The term "cloth" may have been ascribed to the bark material by early visiting European explorers who noticed that the bark was used in native clothing and as a covering material on huts, boats, and other miscellaneous items. The word "paper" is derived from papyrus, but papyrus is a laminated substance with not nearly so much in common with true paper as the hand-beaten bark sheets of the Otomi Indians in Mexico or the native islanders of Java and the remote islands in the Pacific.

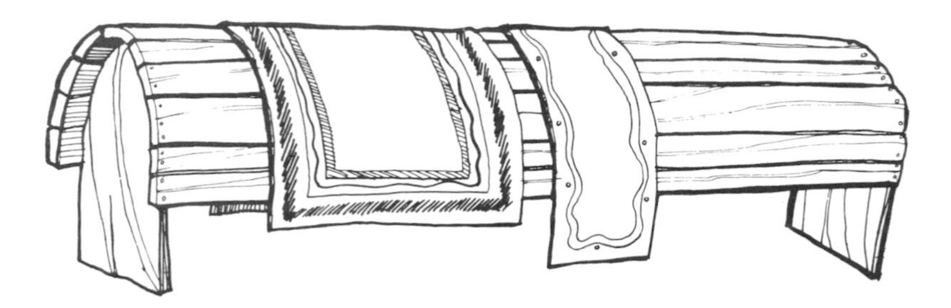

The making of tapa cloth involves hammering the thin bark as it is draped over a wooden support fashioned of thin slats of lumber.

A stamping machine of the fifteenth century. Trip hammers like this one were quite common in parts of Europe and were activated by water as it played down on the paddle wheel.

In Europe, the trip-hammer was used to beat the cotton rags into a pulpy substance by constant pounding. Oliver Bayldon has freely translated a papermaker's Latin poem from the seventeenth century that in part reads:

"The blocks begin their hobnail dance,
and rising, falling, tramp in rows,
stumping on like thunder butts
that crash and crush, and rip and rend
until foundations tremble and sway
buffeting the air around.
More than Etna's torrid groans
when anvils bray for brazen war
to trample out the foot-poised charge:
when bellied drums repeat their call
to echo through the hills and woods
on claustrophobic banks of sky.
The tramping goes on, merciless
till cloth is mashed to milky froth,
but then another pounding yet
which stamps with harder, blunter nails
and even finer beats the threads
until they can be used and only then."

The poem is highly flavored with the kind of activity that is so essential to good paperbeating. Fortunately, the papermaker as a serious hobbyist need not resort to the arduous methods described here. There exists, in the form of the electric blender, an ideal machine that will adequately shred vegetable and plant fiber suitable for the papermaker's need.

An electric blender with several speeds is used to shred raw fiber transforming it into pulp suitable for forming into sheets of paper. The pulp may also be poured over forms.

Use of the Electric Blender

The electric blender cuts plant fiber into small strands capable of being formed into highly useful sheets of interesting papers. The blender must be a well-made one and have several speeds to accommodate various grades of material. Place the blender on a flat, dry surface in clear view. This will enable you to see the pulp change in its appearance and consistency when blended.

Have enough raw material prepared beforehand and place it near the work area for use in the blending process. You may wish to start with paper, plant, or vegetable material. Waste paper will break down quickly, provided it has been cut into ½ inch squares and left to soak for at least two hours. Construction paper, kraft paper, stationary, and watercolor paper are all suitable. Newsprint should be used only as an additive to give bulk to other papers because it is extremely weak when used by itself. Remove all plastic coating, staples, strings, etc. from waste paper lest they injure the cutting blade or bring the motor to a sudden halt.

The density and inherent strength of the paper or plant will determine how much is to be placed into the blender and how long it must blend. Fortunately, this matter is not critical, provided the material is not left to blend indiscriminately. Place approximately two to three ounces of condensed wet pulp into the jar. Add water and pigment if you wish to within two inches of the brim.

COLORING THE PULP

If color is desired, you may want to start with ordinary food coloring, Procion fabric dye, acrylic paint, Rit fabric dyes (which are available in grocery stores), inks or previously blended pulp. Whichever you choose, avoid dyes that are toxic.

There are many coloring additives which are harmless and will give new and electrifying colors to your paper. Try Dr. Martin's Dyes for intense color brightness. These dyes can be found in most artist supply stores. Use one teaspoon of the dye per 12 ounces of compacted wet pulp. The results of this dye will pleasantly surprise you.

Liquitex acrylic colors in tubes or jars will also make your paper more colorful. Again, experiment to find the most attractive results and keep notes in the journal that should always be near your work space.

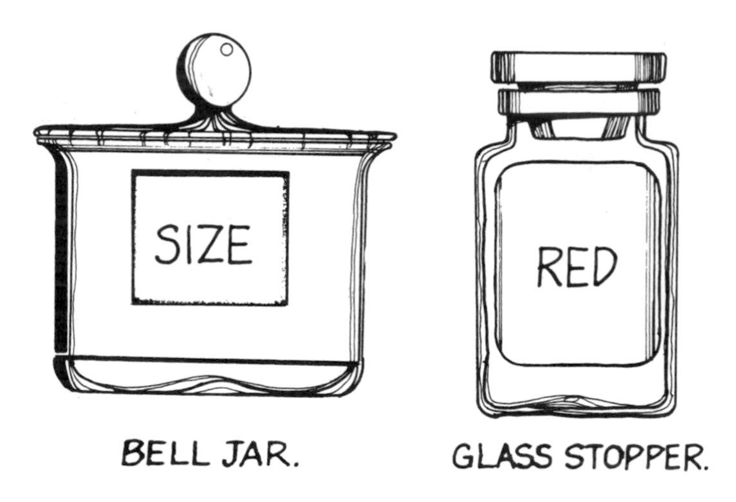

Coloring and sizing material should be stored in clearly marked jars with tight-fitting lids.

SIZING

Make several sheets of paper from beginning to end before sizing is to be considered. Unsized paper evinces the most natural finish and is paper in its purest form. If the finished sheet is intended to carry aqueous media such as watercolor or ink then sizing should be used to eliminate feathered bleeding of the drawn image. Sizing may be in the form of alum, powdered rosin, corn starch, animal-hide glue, acrylic polymer medium, rice paste, unflavored gelatins, and other starchy substances. Use no more than one teaspoon of powdered rosin per quart of wet pulp mixed with water. Starch powders require that approximately one level tablespoon be used with the same volume. The handiest prepared and premeasured additive is unflavored gelatin in small packets available in grocery stores. One packet per quart of wet stuff is all that is needed. The sizing will be quickly and evenly absorbed. As mentioned earlier, you also have the option of sizing paper once the finished sheets have been allowed to dry. This method is covered below

BLENDING THE CONTENTS

Once all the additives are placed in the pitcher, start the blender at a low speed. Maintain this speed for approximately one minute or until the waste paper or plant material is sufficiently agitated. Proceed to the next highest speed; however, watch closely. If the mixture is rapidly becoming fine then stop the motor and add more pulp. Blend for another 15 to 20 seconds. When the blending process is completed, remove the pulp and place in a jar for storing. If the pulp is too fine, i.e., has the appearance of cloudy water, then too little pulp was used. Properly blended pulp from already existing paper will be creamy with extremely small clusters of fiber. If garden plants are used, the variety in texture and thickness may be great. To this end, each batch will provide a different type of paper. Thick paper requires a longer drying time but is an ideal soft paper. In the collage, it serves as a pleasing compliment to more delicate and lacy materials.

If you want to test the pulp after blending for visual characteristics, pour a small amount into a glass beaker or test tube and shake. Clumps indicate too short a blending time or the use of too much raw material.

Unusual blends of pulp and mixed ingredients such as small leaf parts, short bits of string or thread, add a decorative pattern to the paper that may further interest the artist. Simply shorten the blending time or add ready mixed material during the last 5 to 8 seconds of blending. Consider parsley, confetti, Christmas glitter, cut pine needles, small bird feathers, bits of colored ribbon, or even sawdust. Interesting papers have been made with the lint captured in the filter traps of clothes dryers. Just separate the material and slowly mix with the pulp. You will have paper that is swimming with colored cotton and polyester threads suggesting endless possibilities.

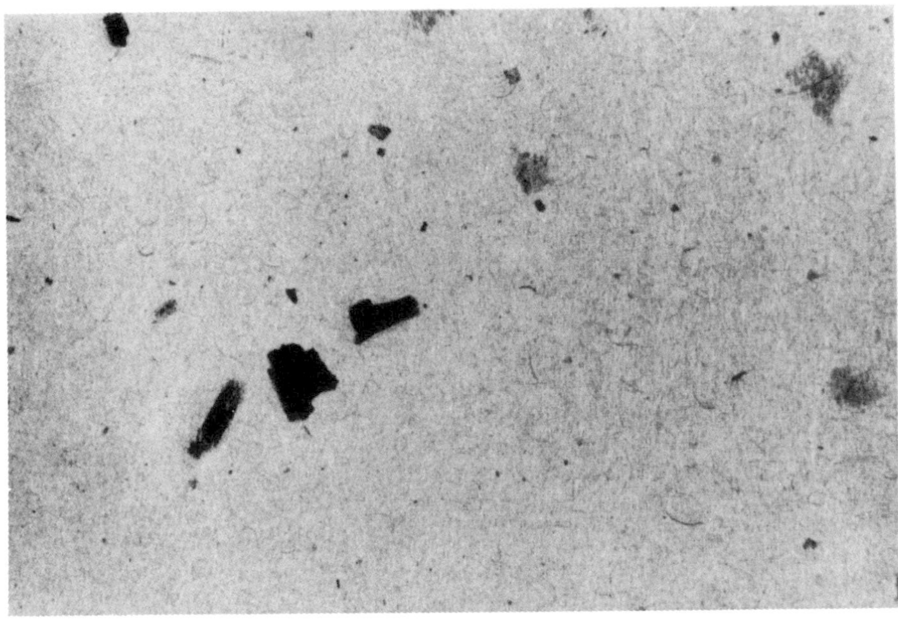

Mottled paper made with small bits of thin bark.

THE VAT

In the early days of papermaking a hollowed-out log or oak wine cask often served as the holding tank for the pulp prior to its forming with the mold. Fortunately, simple tanks may be made or purchased which will function as well if not better. An inexpensive vat can be made out of ½-inch plywood painted with resin to seal the pores of the wood. The walls of the vat should be no less than 8 inches in height. Though it is not necessary to miter the ends of the plywood to form a right angle, it is necessary to fasten them with screws and appropriate glue to guard against spaces where water may seep through.

Apply at least two or three coats of boat resin over the entire surface top and bottom of both sides testing for leaks. The vat should have a flat drain ring installed in one bottom corner. When it is time to drain off the water and pulp, it is much easier to pull the drain plug than to lift the vat to pour out the contents.

Very inexpensive and functional vats may be found in the form of deep trays or "tubs" used by busboys in restaurants. Most cities have restaurant supply stores which carry this indispensable item. They are available in a variety of shapes and sizes and are ideal for use with the small mold. These tubs are a breeze to clean and store and are virtually indestructible.

Some papermakers prefer to have pulp placed into the vat at warm temperatures. This accelerates the dispersion of the fibers and aids in their subsequent formation upon the mold screen. The pulp should be sufficiently agitated with the beater to aid in the hydration (the combining of substances with water) of the plant material. You will also find that the pulp quickly settles to the bottom of the vat and needs to be blended every few sheets to insure proper and even suspension of the pulp.

A "busboy's" tub used for the vat to hold pulp. This item is very inexpensive, durable, and easy to clean. It is ideal for making small (8-by-12-inch) papers.

Basic steps for forming paper. Pulp is added to warm water in the vat. In this illustration paper is to be made using cotton rags

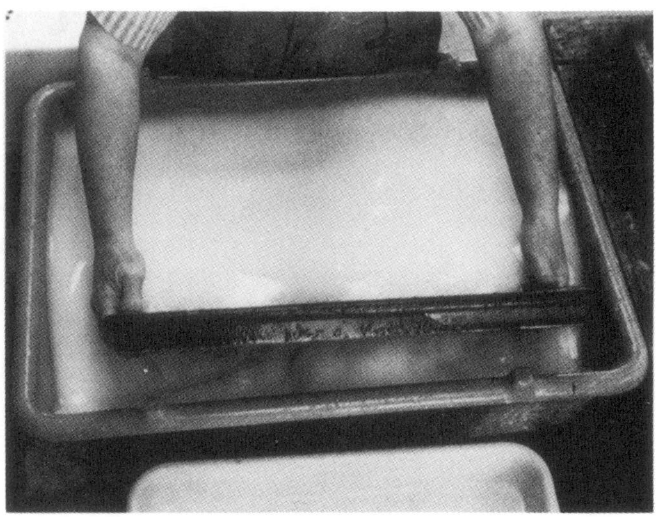

The mold and deckle are lowered into the vat and a scooping motion is used to capture the pulp on top of the mold screen. This is to be performed in one continuous slow motion.

The pulp is agitated by hand. A small electric beater or hand-operated "egg beater" may also be used. Pulp must be sufficiently mixed with water for best results.

The deckle is fitted over the mold and is now ready to be lowered into the vat.

Here, the mold is raised allowing excess pulp and water to drain over the deckle.

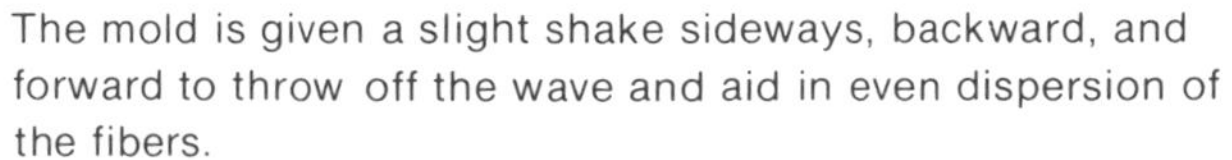

The mold is given a slight shake sideways, backward, and forward to throw off the wave and aid in even dispersion of the fibers.

FORMING THE PAPER

Pour two quarts of concentrated wet pulp into the vat. The pulp may be a combination of vegetable material and recycled paper pulp. Consider at this time the addition of bits of straw, thread, dried weeds, straw flowers or variegated tea leaves. If you are going to make paper using plant and vegetable material, you may wish to eliminate the possibility of your paper developing a mold or even rotting by the addition of one tablespoon of baking soda and one tablespoon of ascorbic acid. To this quantity now add at least one quart of warm water and stir the pulp and warm water together so that the ingredients are evenly blended. You might use an egg beater or an electric beater to facilitate the blending process.

If the pulp appears to be too runny and does not form well (i.e., the fibers do not mat properly) or if the pulp is too thick, then add more pulp or more water to the vat to bring the contents to a desirable consistency. Note that some pulps such as chopped carrots, spinach leaves and maple leaves tend to settle at the bottom of the vat rather quickly, sometimes in the time it takes to couch the sheet sediment has formed on the bottom of the vat making it necessary to stir in between the forming process. Fibers must always be in suspension for a well-made sheet. It will not take long to learn when to add more pulp to the vat in order to replenish the stock. If you find the original three quarts to be too shallow in the vat, then add another quart of pulp and approximately one-half quart of warm water and so forth until your vat is filled to the point where the mold and deckle can be totally immersed before drawing the vat upwards. Fit the deckle (if a deckle is to be used) over the mold, making sure the fit is even and snug. Firmly grasp the two together and position the mold vertically above the vat on the far side away from the body. Slowly lower the mold and deckle into the pulp in the same vertical position and begin to level the mold out so that it is approximately 3 to 4 inches below the surface of the suspended pulp. Without breaking the slow, rhythmic motion of the scooping out process, bring the mold upward and, when the mold and deckle are fully clear of the pulp in the vat, give it a slight shake sideways, backward and forward. This will even out the pulp, help in dispersion of fibers in the water and accelerate drainage from the sides of the mold. This procedure, in professional papermaking practice, is termed "throwing off the wave" and is a most important routine in the making of the handmade sheet. At this point the fibers interlace and become somewhat fixed determining the future complexion of the overall sheet.

A thin stratum of paper is what remains on the surface of the mold before it is tilted sideways to facilitate draining.

If a two-person operation is employed, then the "vatman" passes the mold and waterleaf to the "coucher" who remove the deckle.

The deckle may be removed at this point or after the paper has briefly been allowed to drain.

Tilt the mold to a 45-degree angle and allow it to drain for at least 10 seconds. Remove the deckle. Turn the mold upside down and place one end on the felt. In a continuous, slow, rolling motion press the mold and paper onto the felt exerting even pressure throughout. Raise the mold; the sheet will have been transferred to the felt. Place another felt on top of the sheet and repeat the process. Check the consistency of the pulp and water in the vat every few sheets and add condensed pulp as necessary.

Tilt the mold and allow to drain at least 10 seconds. The newly formed sheet will not fall off the screen.

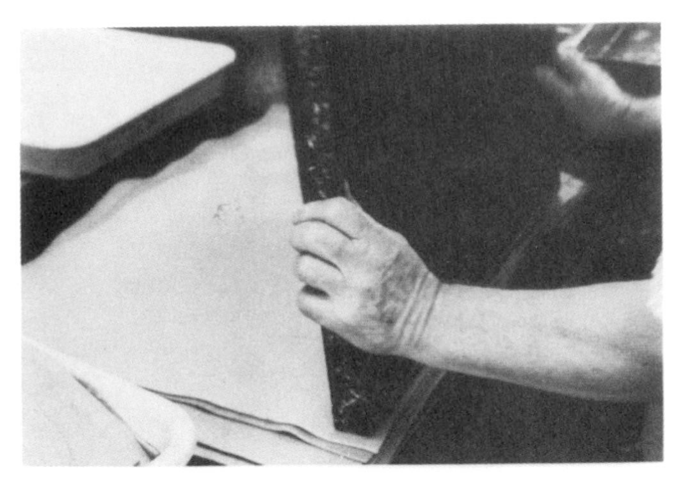

Side view just prior to couching.

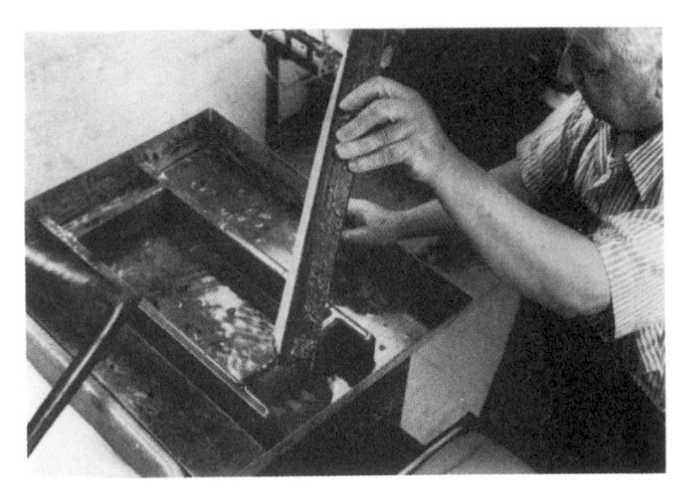

Side view of the draining process.

In a rolling motion the paper is transferred from the mold to the felt.

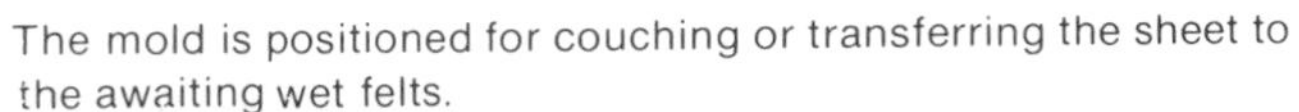

The mold is positioned for couching or transferring the sheet to the awaiting wet felts.

The right hand gently presses down during the rolling motion, while the left starts to lift the top edge of the mold off the felt.

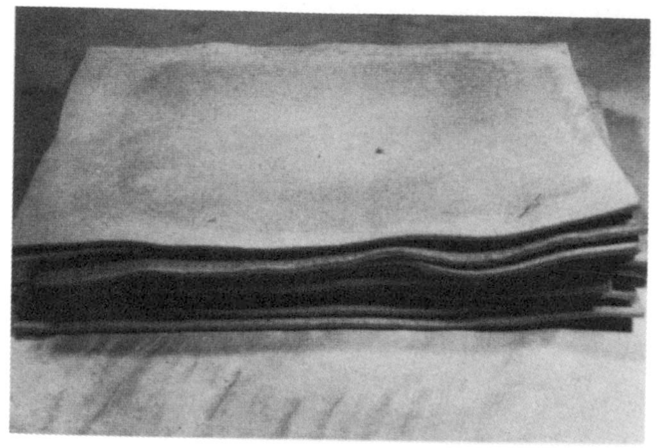

A post of felts and alternating sheets of newly formed paper.

The mold is brought to the surface revealing a shimmering sheet of paper- to-be.

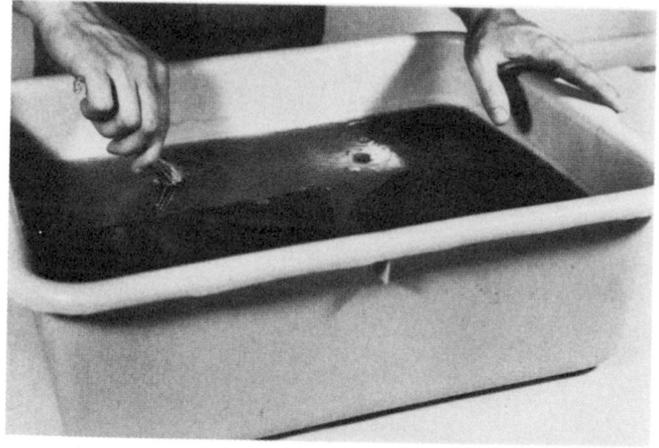

A simpler use of mold and deckle in the forming of natural fibers such as jute, iris, or cattail. Cattail is being formed here. Gently stir to sufficiently agitate the pulp.

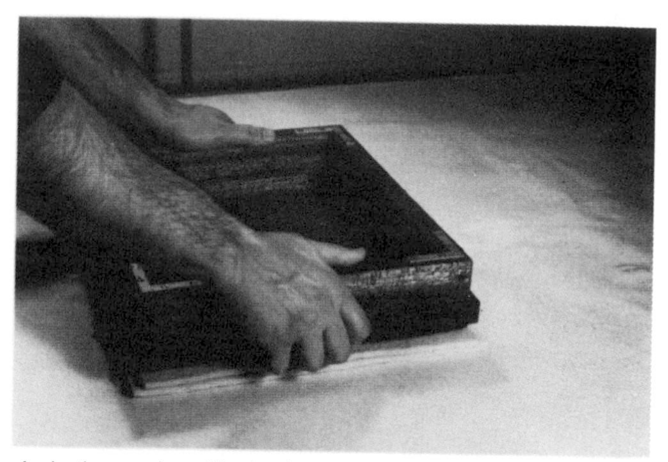

As in the previous illustrations the mold is couched onto the awaiting felts following the standard procedure.

The vat is of the tub type and filled with the long tendrils of the cattail plant. The mold and deckle are lowered into the pulp allowing the fibers to swirl over the surface.

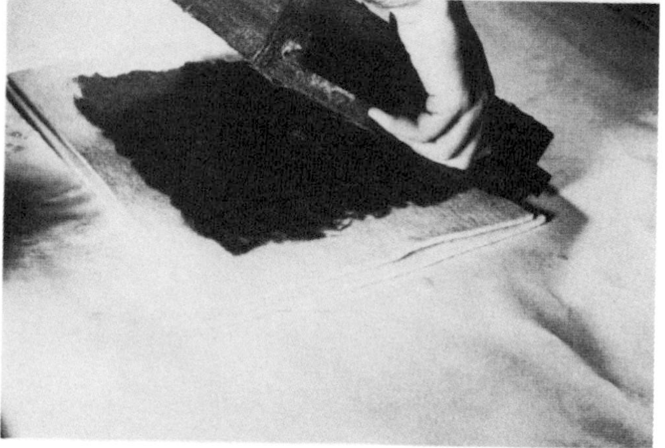

The mold is raised leaving the waterleaf on the felt.

PRESSING, DRYING, AND SEPARATION

If freshly made sheets of paper are not to remain on the mold to dry, they must then be couched onto the awaiting felts. After a post (pile) of felts has been completed, it is transferred to improvised or ready-made pressing equipment to further remove excess moisture. Pressing will help to keep the individual sheets from curling and rippling. After pressing out as much water as possible, remove the post from the press.

You must now separate the sheets, each one on its blanket, and lay them out to dry. In fine weather they may be dried out of doors. Indoor drying can be done by placing the sheets on open-slatted racking placed near a source of low-temperature heat.

The dry sheets are removed from the rack and inspected for undesirable flaws which may be removed with tweezers. Holding the dry sheet to a source of strong light will show you if the sheets have been made with an even or irregular thickness. The thinner parts of the sheet will be more translucent. If this characteristic is repeated too many times, you may want to review the section on how to raise the mold from the vat so that the pulp rests evenly on the surface of the screen.

Once dry, the sheet is examined for finish and texture. Notice the small pieces of dry plant material remaining in the paper. This will impart to your paper a unique quality that most cotton rags will not do.

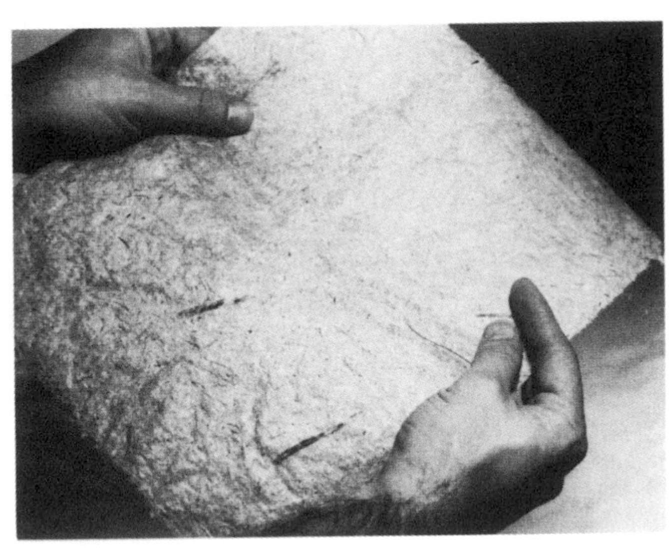

The paper is examined under strong light for undesirable flaws.

The Felts

The felts must be well-wetted blankets, compressed felt composed of particles, blotter paper, or any material that is absorbent and has a smooth surface. Wet felts prevent the fibers within the sheet from tearing upon transfer. Textured towels will leave an impression on the finished sheet and you may find this feature an area to explore. The paper, if not made smooth by use of extreme pressure, can take on the fàint appearance of the blanket or blotter upon which it has been placed to dry.

The wet blankets or felts are larger than the individual sheets by at least 2 inches all the way around. This makes the couching process simpler, allows for a less than exact placement of the waterleaf (the newly made, wet, unsized sheet), and aids the absorption of water. The felts are placed in an over-sized stainless steel or wooden tray about 1 inch in height so that the tray collects water as it drains off the material. It is best to use 7 to 10 felts to make as few as 5 to 8 sheets of paper until you develop experience in determining how much pulp is necessary to make a specific number of sheets. Felts should be dried after each papermaking session in order to prevent mold and mildew.

The felts or blankets must be wet in order to achieve a smooth transfer of paper. Felts are usually made of highly compressed felting and are ideal for papermaking.

The Press

A number of ingenious devices and means have been developed over the last two millenia to press dampened paper between felts with sufficient pressure to remove as much water as is possible. In the Orient and Near East long poles with rocks on the ends serve as counterbalancing weights to drive out excessive moisture from the post. Recently, one rather anxious papermaker was known to have placed a piece of thick plywood on top of a low post of paper. He then slowly drove his car over the surface of his efforts. Fortunately, there exist better means of pressing paper in order to consolidate the fibers for ease of handling and drying.

An example of paper that has been left to dry on a coarse towel which left a textured impression.

An early drying press using a log and several stones. This type of crude press originated in Japan and is still in use there today.

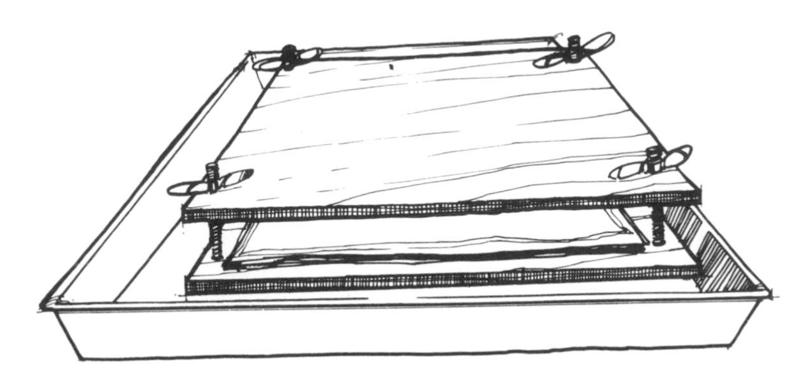

A simple, yet effective, method of squeezing the newly formed sheet to rid the post of excessive water.

One simple pressing device can be constructed of plywood, carriage bolts and wing nuts. The procedure is as follows: take two pieces of 1-inch thick plywood cut to a size at least 2 inches larger than the planned post of felts and paper. In each corner of the plywood drill a hole to accommodate a 4-to-8-inch carriage bolt with a galvanized finish. Once the post has been placed in between the pieces of wood, the wing nuts are screwed down, forcing the plywood to clamp down upon the post. Not a great deal of pressure will result, but enough to meet the usual requirements.

Pressing can also be achieved by use of a bookbinder's screw press. These presses are capable of exerting much pressure, and provided they are placed in a pan or tub, this phase of clamping and squeezing the paper will be easily completed.

If the above items are not available, pressing can be done with a common household rolling pin or by placing heavy stones on top of the board which caps off the post.

Whatever equipment is used, a wooden board must be placed on top of the pile before squeezing. The board should be at least 1 inch thick and oak, teak, mahogany or other wood that will resist warpage is preferable.

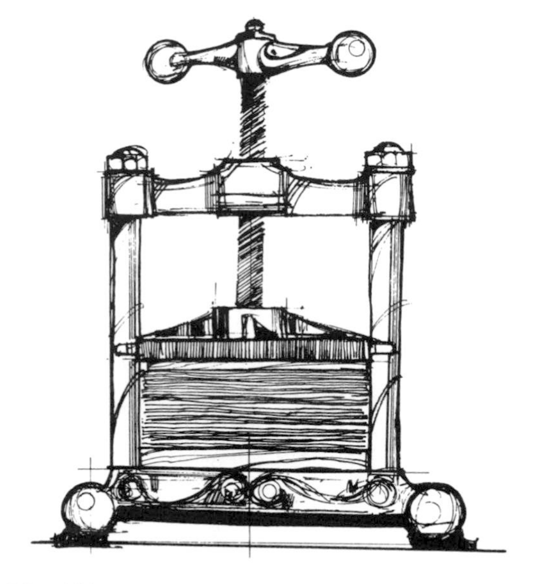

A bookbinder's screw press. An ideal tool for the papermaker.

A person handy with tools may build a press similar to the one pictured here. It is not difficult and should provide years of reliable service.

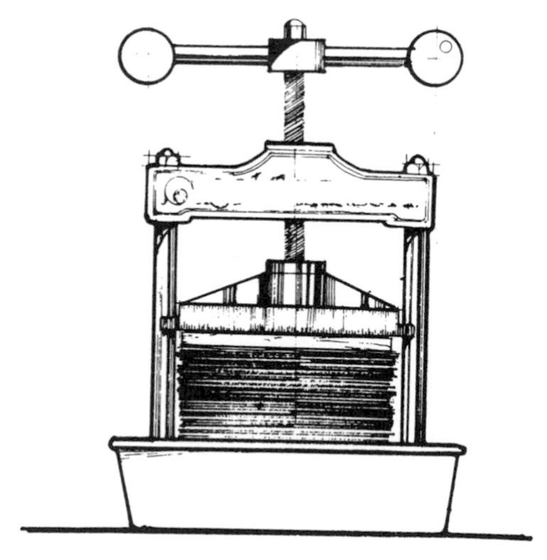

The press is placed in a large plastic or stainless steel tray to contain the water as the sheets are pressed.

Sheet Separation

When the sheets have dried to the touch on the felts, they need to be removed and placed in a pile. The pile is once again subjected to the screw press until fully dry. This last step is a subjective one calling upon individual experience. Dry to the touch does not mean that the paper is terminally dry. Paper at this stage is slightly absorbent and suitable for printing. If watercolor or writing ink is to be used on the surface, the paper will first need to be sized. Paper may still contain traces of water and will need further curing sometimes lasting a week or more. Again, the drying of paper is subject to many variables such as thickness of paper, plant composition, prevailing air temperature, etc.

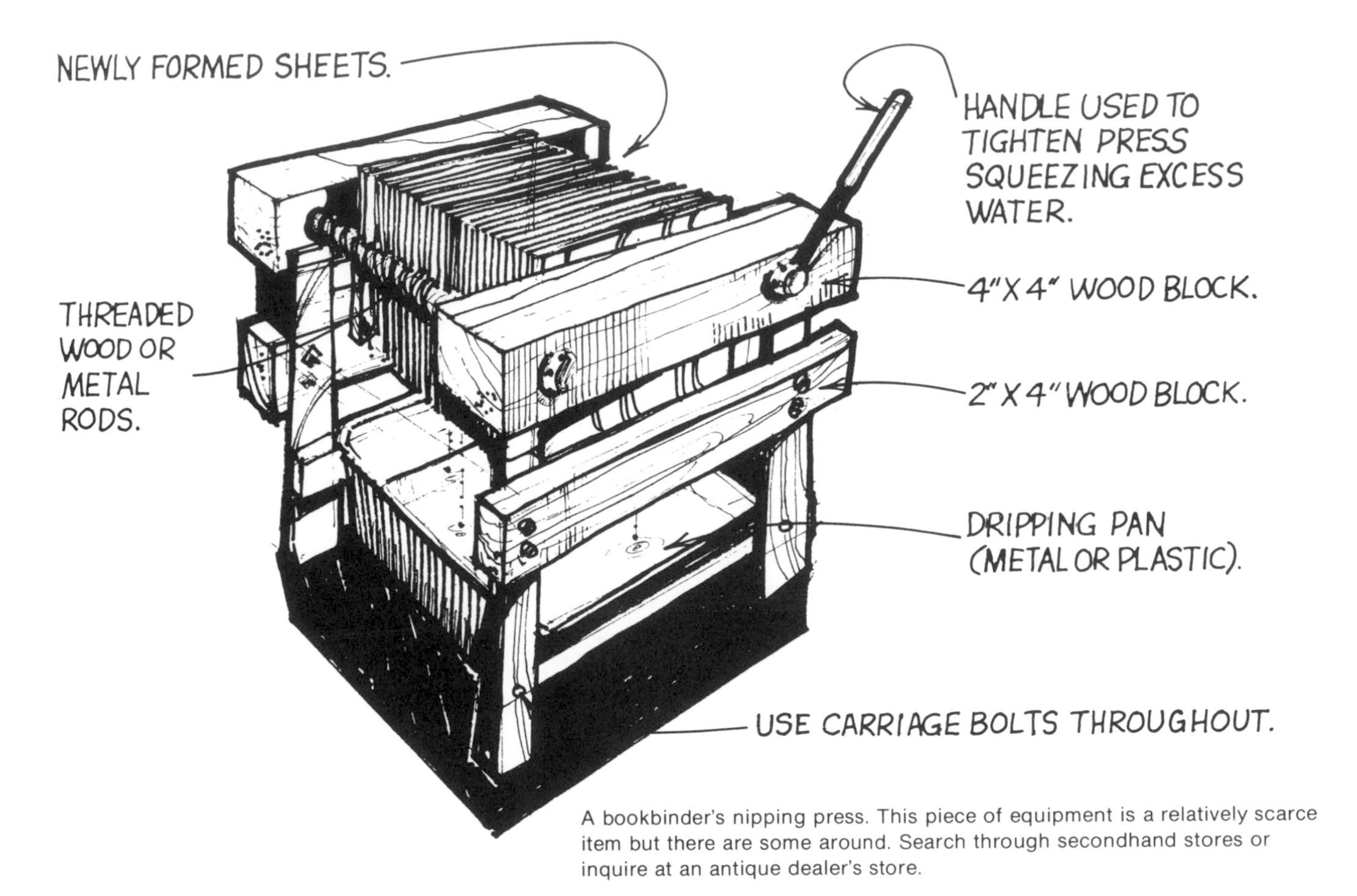

A bookbinder's nipping press. This piece of equipment is a relatively scarce item but there are some around. Search through secondhand stores or inquire at an antique dealer's store.

An electric iron may be used to lightly dry sheets of paper. The iron must be heated at a very low temperature and gently skimmed over the surface of the paper. The iron will also help to flatten small pieces of paper.

SIZING

There are two ways in which the waterleaf or dried paper may be sized. Both methods have advantages and disadvantages. The first method is to add sizing made of uncolored and unflavored gelatin mixed with water to the pulp while it is in the vat. The pulp and water in the vat must be very warm, almost hot. This prevents the gelatin from turning prematurely into a jellied state which would render the sizing useless and create small beads of semihard jelly throughout the vat and its contents. To this premixed recipe of water and gelatin, a small amount of alum or powdered rosin may be added. The measure is approximately ½ teaspoon per quart of premixed gelatin size. The alum or rosin aids absorption and also helps to ward off fungus and mold on the paper which are an anathema to good and successful papermaking.

Powdered gelatin is available in small premeasured packets. Use one packet per quart of hot water. The mixture must not be allowed to cool or it will turn into colorless jello. Pour the liquid-gel sizing and alum into the tub containing the warm pulp and gently stir until all is properly blended. Form and couch the paper as described earlier. Once the paper is removed from the felts, wash the felts with warm water and a small amount of laundry detergent to eliminate the size. This will also help to preserve the felts and keep them smelling fresh.

The second method of sizing paper is to immerse the sheets in a trough of size about two weeks after the paper has thoroughly cured. The reason for the long wait is that if not totally dry the paper may disintegrate in the trough of water and size.

The first few times, gently place an individual sheet in the premixed solution described above and slowly depress the paper to the bottom of the tub. Build a pile of about 8 to 10 sheets and, placing the pile between boards, insert it into the press. A quick squeeze of a minute or so is all that is required.

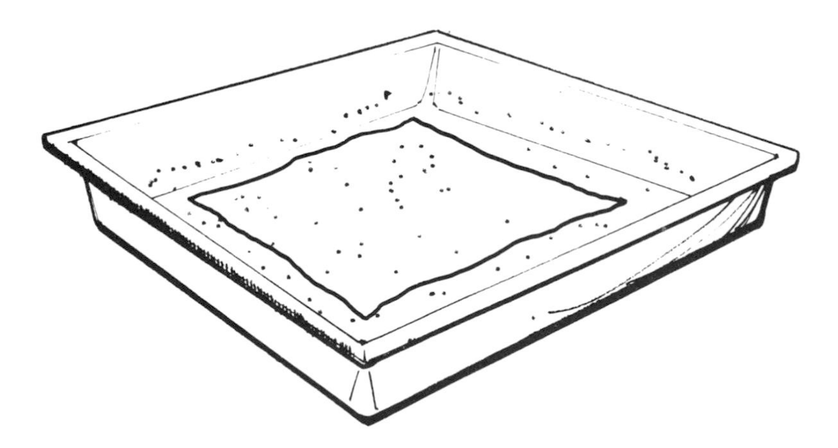

The sheet is immersed into the sizing solution for a brief moment and then removed to be placed into the press for further squeezing. The size must be very warm so that the gelatin does not become jellied.

This blots the paper and insures absorption of the size. Remove the paper and felts from the press and place them on the rack or table to dry. As above, wash the felts in the prescribed manner.

There exists many other ways of sizing paper using less readily available chemicals. If powdered rosin is in your reach, you may wish to experiment with this in your pulp mixture, determining what amounts are best for your individual use. The above mentioned gelatin is nontoxic, expendable, and extremely reliable. For the humble needs of the handmade paper hobbyist, it is perfectly acceptable. Exotic formulas are not necessary here. They are oftentimes troublesome and can defeat the simple objective of making paper useful for aqueous media.

As with all other steps involved, it is most important that you get going as quickly as possible in order to develop a level of interest and excitement that will give further momentum to your artistic papers. There is plenty of time to make improvements as experience develops and the need arises.

In the early stages of making paper by hand it is suggested that you use whatever resources are available and within reach. This, coupled with native intelligence, will bring about results that can be analyzed at a future time when reflection will provide new and improved ways to make your personal papermaking equipment more workable.

5. Projects

POURING PULP OVER A THREE-DIMENSIONAL FORM

Materials

Prepared pulp, 1 quart
Pyrex baking dish (small or medium size) or prepared mold with attached screen
Spoon, small
Vegetable spray (nonstick aerosol)
Knife
Flat wood spoon or spatula
Three-dimensional forms (shells, buttons, medium size seeds or pods, smooth stones, rope, yarn, insects, leaves—avoid anything metallic in composition)

An example of what unusual configurations paper can assume when it is poured over a three-dimensional form and allowed to dry. The material for the form used here was jute twine and coarse yarn with tufts of wool. The pulp can be an extremely faithful material and is capable of revealing every nuance of the mold.

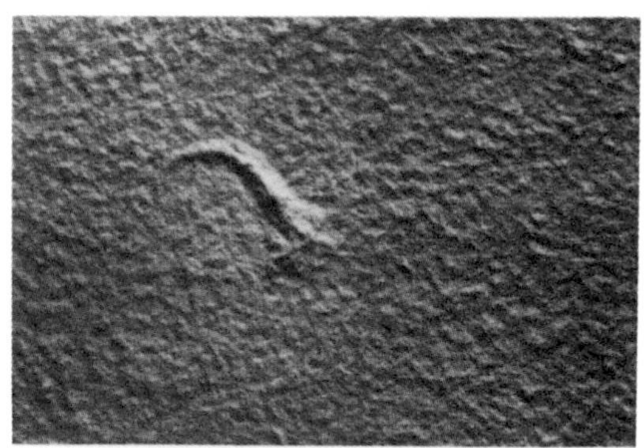

The surface of this form is gently disrupted by the presence of a piece of string underneath the paper.

Procedure

The objective in this project is to create a form of paper sculpture or relief by pouring liquid paper of any color or texture over a low-relief shape or shapes. When the paper is dry the positive object is gently removed revealing a negative impression in the paper. The impression will be quite faithful and should reveal most or all of the nuances of the three-dimensional object. The molded paper form will be lightweight and relatively strong given that the fibers have interlaced themselves without aid of suction as in vat formed paper.

Clear a space on the work table and wipe free any dirt, colorants or stray material that may interfere with the project. Place the forms into the dish or onto the screen if you are using the mold. (Note: the mold will allow excess water to drain faster, thereby hastening the drying process.) Arrange the shapes so that they are visually pleasing and well composed. Any object that is highly irregular or too complex in form may prove to be too difficult for initial experimentation. Lightly spray the objects with vegetable spray. This will facilitate the separation of mold from paper once the sculpture is dry. Replace the objects in their prearranged positions and prepare to pour.

Blend and sufficiently agitate the pulp prior to pouring. This will break up clumps and allow the fibers to flow smoothly and evenly. Be sure to use warm water mixed in with the wet pulp, as this assists the pulp in its dispersion. Once the forms have been set, slowly pour the pulp into the dish or onto the screen being careful not to disturb the objects in place. Pour approximately one-third of the total amount to be used, then set the pitcher aside. Using the wooden spoon or rubber spatula, gently spread the pulp around the forms if the pulp has not already spread evenly into these areas. If the pulp is too thick and will not respond to leveling, then the process of shredding or beating in the blender should be repeated with the addition of more water or less pulp.

Once the pulp has somewhat leveled, continue to pour more material over the first layer. This first layer will help to anchor the forms at their base and less movement is likely to occur. Use the spatula and spoon to spread the top layer and even out the surface. Keep in mind that each successive layer or "stratum" of paper may be of a different color. If the shapes penetrate the paper by an inch or so, the effect can be dazzling. As the colored pulp dries it will lose some of its glistening brightness and will settle into a quieter shade or tone of its original brilliance. Nonetheless, the overall effect will have a well-integrated appearance.

The shapes may be covered with the pulp or allowed to penetrate the top of the pulp. Place the entire project to one side and allow it to dry slowly. The drying permits the fibers to thoroughly knit themselves around the form and to contract with the evaporation of water. The workable mixture has a water content of approximately 94 percent and pulp content of 6 percent. Once the paper dries, the figures are reversed, 94-percent pulp, 6-percent water. Hastening the drying process can be achieved by placing the dish and contents near a source of heat such as a heater vent (not so close that it is potentially combustible) or simply leaving it in a gas oven with only the pilot light to keep the interior of the oven warm.

When the paper appears to have thoroughly dried, it may then be removed from its form. Gently wedge a thin tool such as a butter knife or shim down the side of the dried paper and lightly pull the blade around the top part of the form. This will sever any strands that have overlapped the top of the dish or mold and will simplify the lifting procedure. Be sure to cut around the entire form. The paper sculpture can now be lifted and removed with little or no difficulty.

Then, invert the cast and gently remove the three-dimensional objects. Disturb as few of the delicate strands of paper as possible. Small parts can be removed with tweezers. Inspect the sculpture for foreign material. Undesired spots of inadvertent color can be bleached with household Chlorox diluted with 50-percent water.

EMBEDDING OBJECTS INTO PAPER

Materials

Prepared pulp 1 to 2 quarts (depending on size of tray to be used)
Tray with sides at least 1½ inches high (plastic, glass or stainless steel)
Spatula or wooden spoon
Objects such as acorns, leaves, wheat stalks, marbles, fish skeletons, coins, colored yarn, straw, small animal bones, thistles, plastic shapes.
Tweezers
Teaspoon

Procedure

By embedding miscellaneous objects into paper the artist fuses both object and paper to create an integrated, well-designed form. Embedding is a simple process and is ideal for children in that the wet paper is thick and responds to implantation of practically any small object. As the paper dries, there is a tendency for the individual strands of the paper to shrink around the three-dimensional shape thereby holding it in place. If the object is not too deeply rooted in the paper, it may be removed revealing characteristics of the shape as discussed in the first project.

Using the electric blender, prepare pulp for pouring. After 30 seconds of blending, remove the pitcher and pour contents into a square or round tray. Use the spatula to spread the material evenly while working the pulp into the corners of the tray. Fill the tray to within ⅛ inch of the top lip. Pour the balance of the pulp back into a storage jar. Note that the pulp is a thick creamy substance. As mentioned earlier, the paper sets up as the water evaporates. In this project the pulp must dry for a while before anything may be embedded. The pulp, unless unusually thick, does not have enough strength at this point to hold anything heavier than a small coin.

Long wisps of fine plant stems were immersed or embedded into the pulp then left to dry. The pulp will shrink slightly around the object holding it firmly in place.

A thin leaf was imbedded into the pulp. After an hour's drying, a small amount of pulp was poured over the surface, thereby helping to hold the leaf in its place.

After several hours examine the paper by lightly probing it with a toothpick. Place the toothpick on end in the pulp. Is it able to stand upright? If so, proceed to arrange and place the shapes into the substance. Allow some of the surface of the object to reveal itself through the paper. If coins, buttons, leaves, etc. do not settle by their own weight gently press the objects into the paper. They should stay in this position until dry.

Paper will often "leech" color from foreign objects when the pulp is wet. Be prepared to know beforehand those colors which are not fast or permanent. The effect of color bleeding is often very attractive and the results may be a pleasant surprise.

The drying period, depending on the source of heat, may take anywhere from 3 days to a week. The thickness of the pulp determines, in part, the length of drying time. During the spring and summer months place the piece near a window for exposure to the sun. The drying also lowers the level of the pulp as the water evaporates. Considerable shrinkage should be expected. This makes the paper more manageable and reveals the contour of the forms embedded into its surface.

If for any reason the pulp is thin after the drying period of several hours, lightly spoon more pulp onto the thin spot. The fibers will fuse and no noticeable aftereffects will be seen. Use tweezers to remove any embedded objects that may be undesirable. Some of the most interesting projects created using the embedding technique call for the removal of some shapes while leaving others in the paper. This is one reason why this technique is so popular with beginners. Random selections are spontaneous and no guidance is necessary.

Once the project is completely dry it may be hung on a wall or framed within a shadow box created to show it off to its best advantage while protecting it from being disturbed. In the classroom the projects can be placed on the display board for a mosaic effect and for others to see.

LAYERING

Materials

Prepared pulp, 2 quarts for each project
Tray, approximately 8-by-14 inches (plastic, glass, or stainless steel)
or
Mold with plastic, nylon, or wire mesh
Thin materials to be placed between layers (leaves, doilies, butterflies)

Procedure

In the Orient, decorative paper is commonly used by artists in a variety of ways. Thin, found objects from nature are placed between two layers of translucent paper and set outside until dry. The paper reveals soft and detailed textures and colors of the objects sandwiched therein. Decorative items such as colored thread, leaf skeletons, cutouts of paper, stamps, etc. can lead to very interesting possibilities. If two sheets of colored paper are desired, then pour the two pulps into two separate vats and use two molds. This procedure will help you avoid the pitfalls of having pulp inadvertently become mixed with one another.

One of the papers should have a diaphanous, almost gossamer finish and texture. The thinnest layer of paper will do. Too much thickness in the newly formed sheet obscures the objects being covered.

The layering method. Once a sheet has been formed, a leaf or some other flat object is placed on top. Allow to set for several hours, then place another formed sheet on top. This way the object is "sandwiched" between two thin sheets of paper.

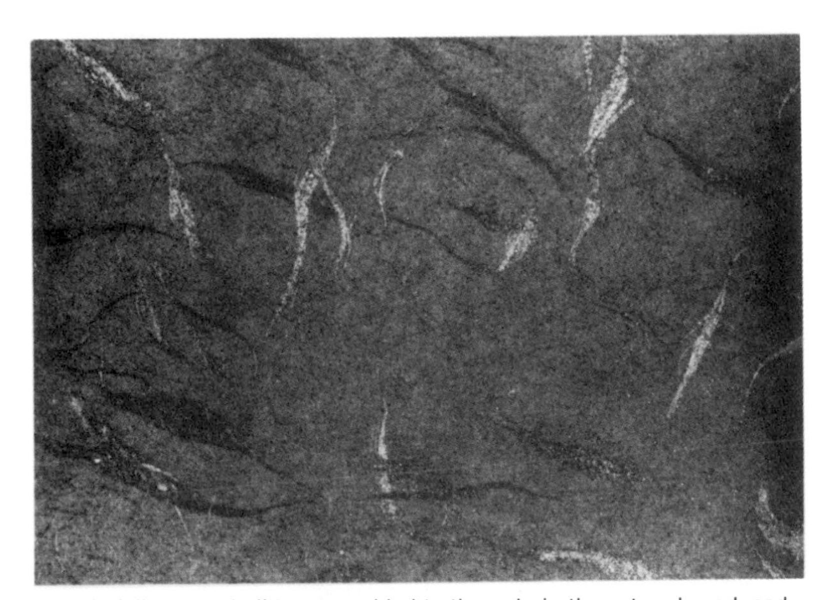

Metallic foil-covered glitter was added to the pulp in the vat and produced this sparkling sheet. This paper has been sandwiched with paper containing long strands of cattail fiber.

Determine the overall size of the finished sheet. If a screen or tray does not exist to your specifications construct one following the procedures illustrated in the section on The Mold and Deckle in Chapter 3. It is recommended that first experiments be done using the wire mold. This will facilitate the drainage of excess water and accelerate the drying time. The glass or plastic trays will suffice in a pinch but expect at least a three-day drying period in warm weather.

The first layer of liquid paper is poured onto the screen and, with the aid of the spatula, spread it into the remote corners of the frame so that an even layer is formed. Have the decorative items to be used ready and place them with care on this first layer. There is no need to hurry. Take time to consider design, composition, texture of the dried works, color, translucency and opacity, thickness, and all the other decisions that go into a work of art.

Once you have decided on the configuration of your idea and its attendant composition, proceed with the pouring of the second layer. As mentioned, this layer may be of a contrasting color achieved by mixing pulps, dyeing the material beforehand, or by sprinkling into the pulp small bits of paper of a confetti nature. Some spices are nothing more than dried, miniature plants and may be used accordingly. Treat them in the same manner as you would confetti paper.

It is a good idea to presoak the material to be sandwiched for two or three hours before setting about creating your work. This will soften any material that is brittle or resistant in nature. Of course, some goods will not respond to soaking and some may deteriorate if soaked too long. If butterflies are to be included, they must be placed in an insect relaxing jar beforehand to allow the wings to unfold and reveal the extraordinary beauty of nature.

The first layer and the materials on top should be couched onto the awaiting felts before the next layer is applied. Struggling to remove the entire work from the mold frame can result in ruin. At best, the matted fiber has begun to interlace but has not knitted sufficiently to withstand rearranging or relocating.

If the second layer has only the merest wisp of fiber, then the transparency of the paper is heightened. The couching of this second sheet on top of the first sheet must be done with a most delicate touch lest the fibers prematurely separate or fall apart. Once in place, a soakened felt is gently placed on top and left overnight to absorb the moisture of the newly cast sheets.

It is not necessary to place the post into the press to hasten drying. In fact, it is better not to do so. The paper will not always withstand the tremendous pressure, and, without calibrated marks on the screw-type press, it is difficult to determine how much pressure is enough.

The paper may dry unevenly even when dried slowly by natural means. However, this form of papermaking is consistent with the attitude that good paper is not always flat, even, and without texture. Your most intriguing works will be the ones that are uniquely individual.

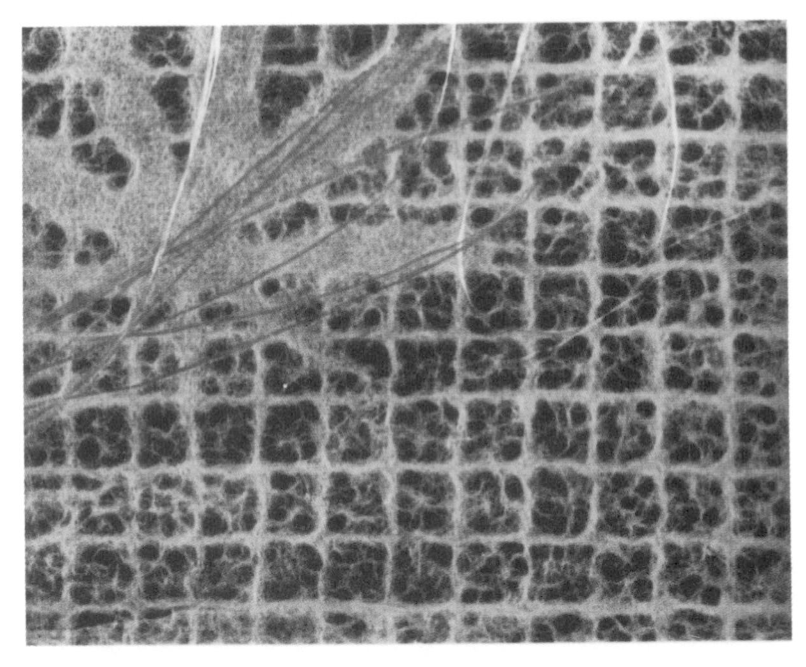

The layering of objects in paper can be a most fascinating manner of working with paper. Japanese decorative paper, pictured here uses this technique to hold bits of nature in suspension with a gossamer window as a protective covering.

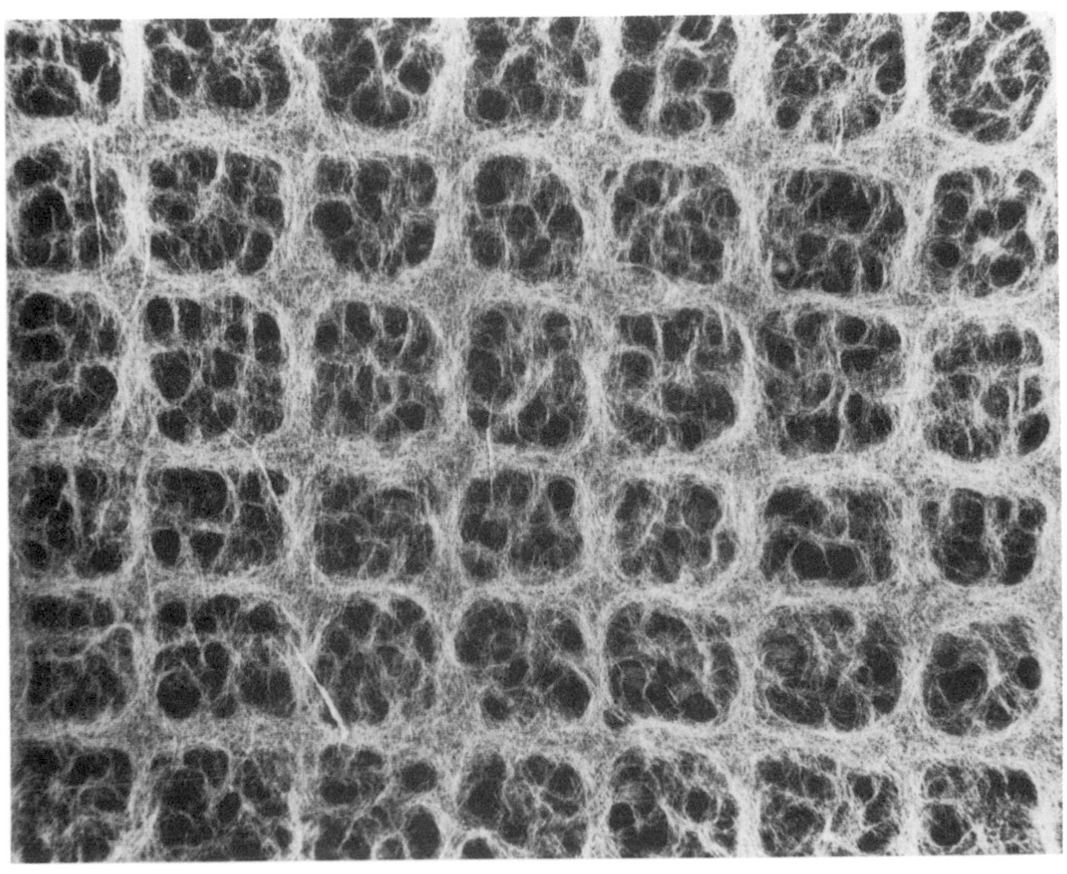

Another example of Japanese decorative paper that is highly porous. Delicate and light, this type of paper reveals an open network and weave making it a good paper for layering technique.

Japanese "kozo" paper reveals the long fibers that swim throughout the finished sheet. It is an ideal paper to use in conjunction with other thin papers.

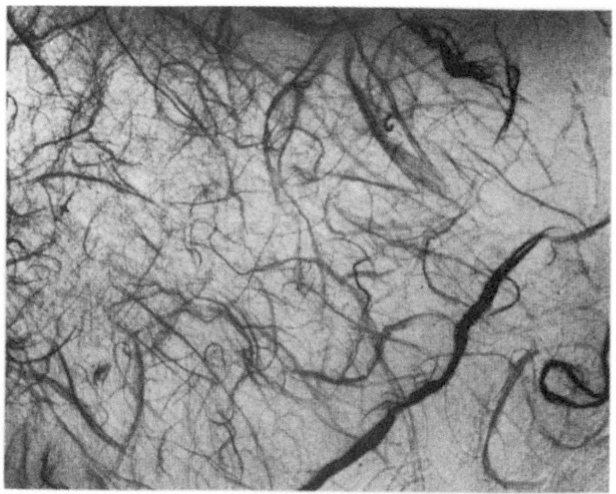

The pulp here is made with very fine and long fibrils of celery and artichoke. The fibrils are mixed with bamboo pulp which has been macerated by the blender until it is almost too runny. The long fibrils help to strengthen the paper once it has dried.

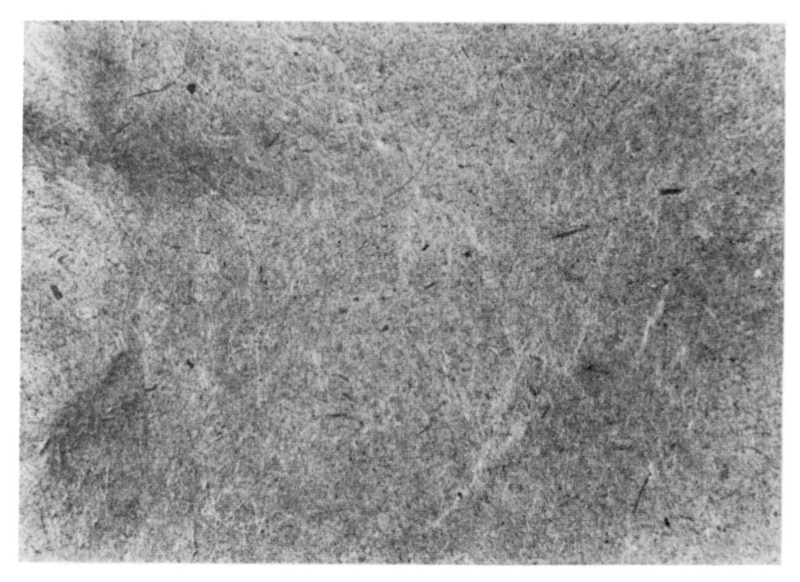

Paper that has been made using gladiola leaves layered with paper made of straw. The finished sheet is rough and similar to oatmeal paper. It is an ideal paper on which to draw using crayons, pastels, or charcoal. This is an unusual artist's paper with no counterpart commercially available anywhere.

An enlarged close-up of paper made with eucalyptus leaves. This paper is very dense and thick and can be used as a base support for other layers of paper to be placed on top.

POURING PULP INTO A FORM (CASTING)

Materials

Prepared pulp, ½ to 1 quart depending on the size of mold
Plastic funnel for directing the flow of the pulp
Spatula
Masking tape
Spoon
Hand drill with 1/16-inch bit
Vegetable spray (nonstick aerosol)
Objects such as shells, desert molds, dried and sealed clay molds, butter molds, exoskeletons of shellfish, etc.

Procedure

Casting with liquid paper reveals to the artist many possibilities. The pulp is similar in consistency to thin plaster but is without the attendant problems of drying and excessive weight. The same mold may be used repeatedly if the artist wishes to create an edition or series of identical pieces. The best molds, whether ready-made or handmade, are those that will not cause excessive adhesion and will allow for an easy release. In any event, the inside of the shell, butter mold, or any other object, can be sprayed with vegetable spray which should not discolor the paper. Experiment with various colors of pulp to find which ones are the most striking or appealing.

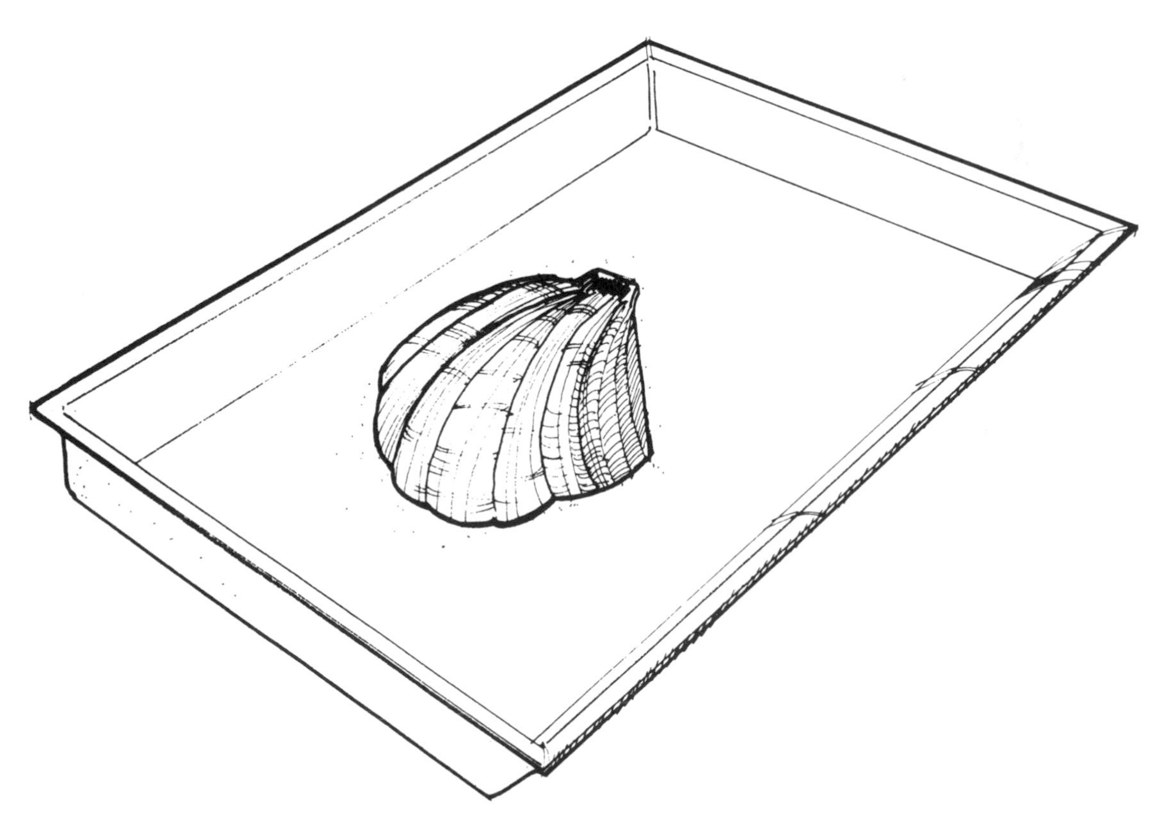

Pour a small amount of pulp over a small object and allow to dry. This procedure is known as casting and can extend the papermaker's craft by allowing the artist to invent new forms with "white art" pulp.

Select an object or two similar to the ones listed above and lightly spray the inner surface with vegetable spray before pouring the pulp. If your form is expendable, drill five or six holes in the bottom to facilitate the draining of water. Before the pulp is placed inside, tape the holes with masking tape on the outside. This will prevent the fine strands from escaping the mold through the holes until the paper has begun to set. As in the other projects, pour slowly and work the pulp into the recesses and contours of the mold. Molds that are complex in shape may prove somewhat difficult. While the paper is in a fluid state it will tend to move easily into tight corners and concavities. However, when dry, it is not easily removed due to the interlocking of the forms. The mold may have to be split by means of a thin chisel or even cracked. If the mold is to be disposed of, this should pose no real difficulty.

Liquid paper flows within the smallest of objects. Interesting directions using small forms such as split walnut shells, seed pods, and rocks with voids and hollow recesses have provided the paper caster with a variety of choices to explore. Clam shells and exoskeletons of crustaceans such as lobster and crab will also tempt the papermaker.

When using found objects it is best to thoroughly clean the form inside and out so that no impurities are transmitted to the paper. A mixture of 1 teaspoon of bleach, several drops of dish-washing detergent, and hot water will clean most forms found in nature. Allow the mold to dry before beginning the casting process.

For fun, children may pour pulp into cookie or pastry molds. Leave the pulp to dry for several days before removing the mold. What remains is a perfect paper sculpture which can be made all the more attractive if color or bits of thread from dryer lint have been mixed with the white paper pulp.

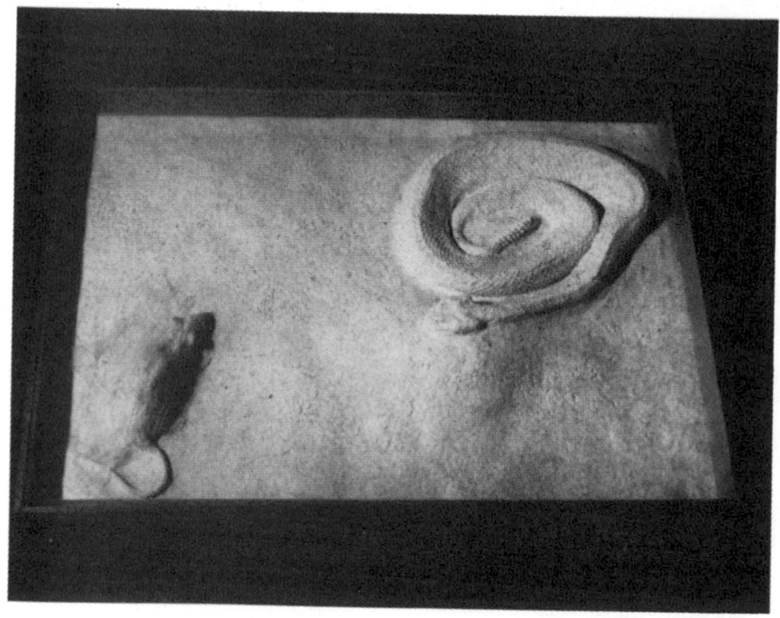

A cast piece by artist Bella Feldman. Untitled.

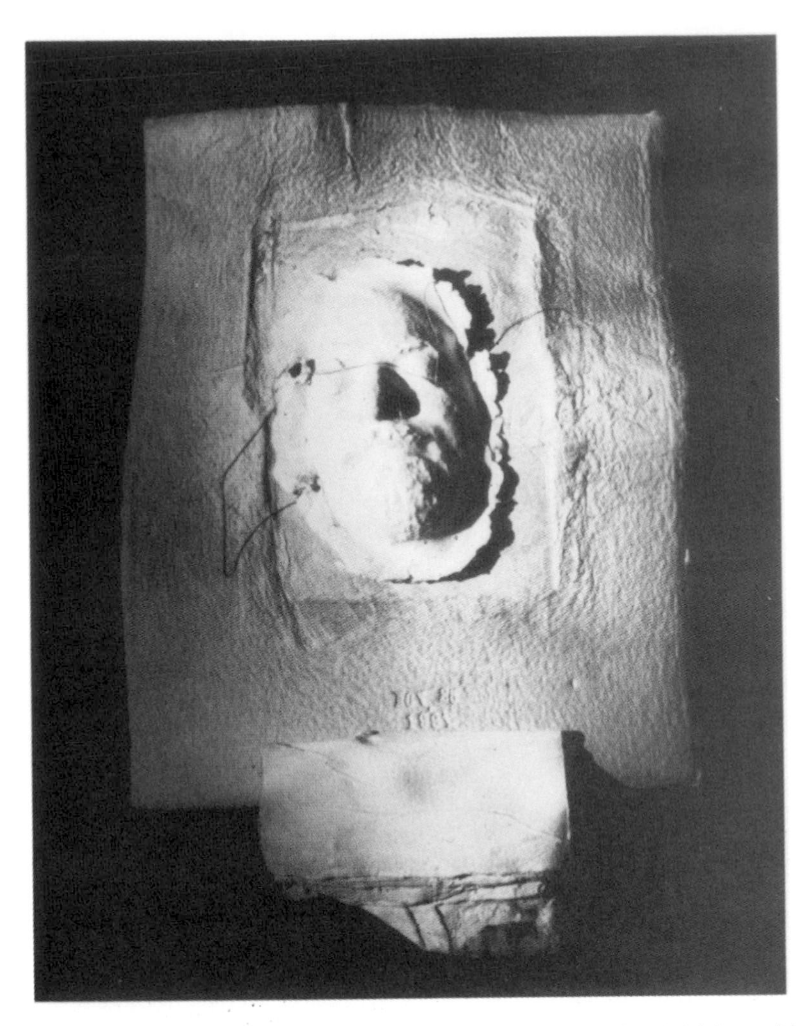

Prophesies of My Future Self 1976-1996. Artist: Harold Paris. Multiple-mold formed sheets, embedments, and mold-formed face.

Box #49. Artist: Charles Hilger. 1976. Hand-formed cotton paper. A subtle blend of various forms of papermaking techniques.

To be sure that the paper dries within a reasonable period of time, place the molds in a gas oven with only the pilot light to provide heat. Molds also may be fashioned out of clay. Coat the thoroughly dried clay interior with vegetable spray and follow the technique described above.

Primordial Dawning. Artist: Vance Studley. A fish skeleton on which was poured two layers of thin white pulp left this revealing fossil-like image.

A close-up showing the pronounced impression of the fish skeleton in the paper. The image is low-relief and a faithful facsimile of the original form.

Untitled. Artist: Margo Studley. A wall hanging is artfully combined with paper casting to give visual richness to this piece.

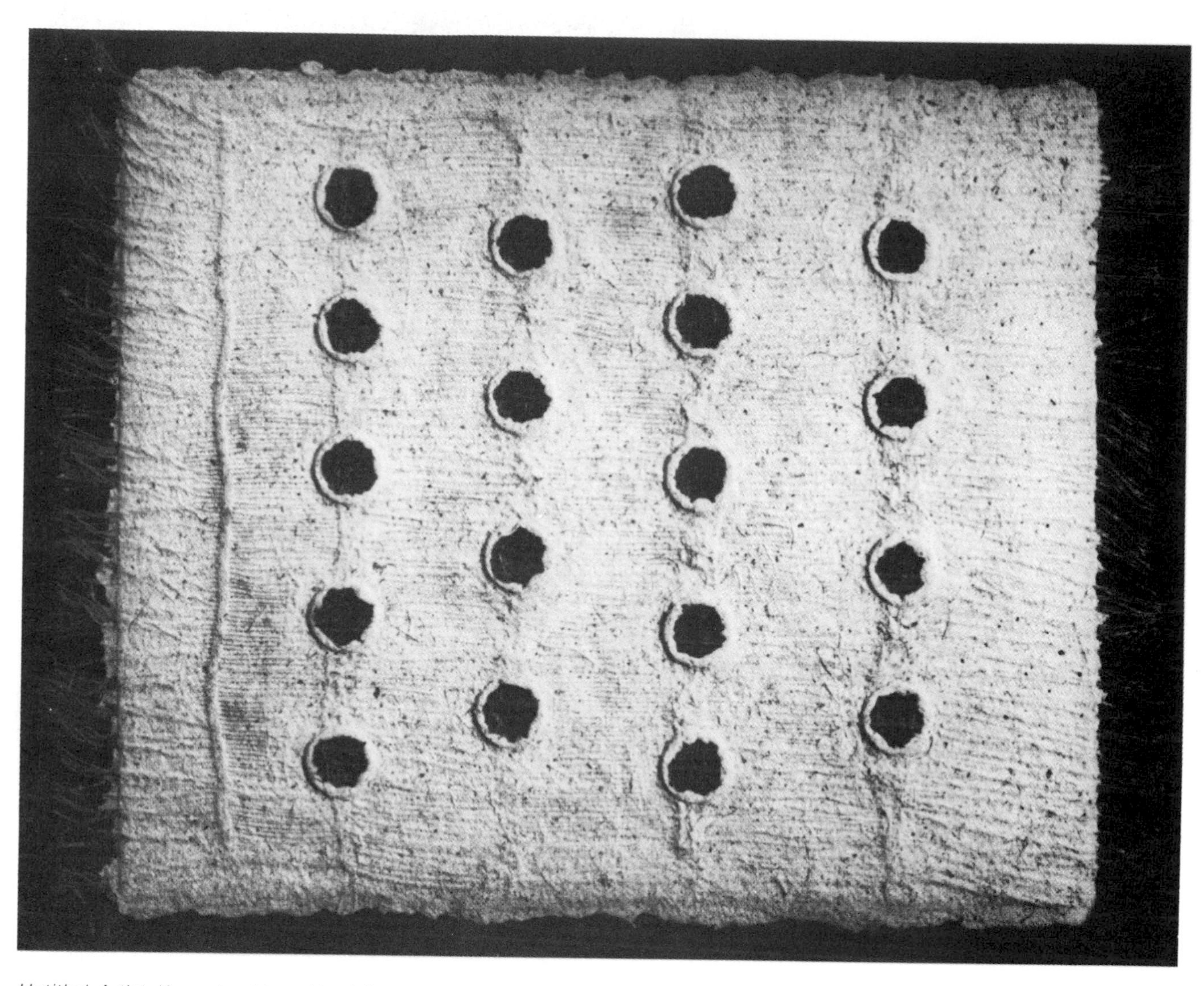

Untitled. Artist: Karen Laubhan. Hand-formed cotton paper, weaving, Plexiglas discs, coconut fibers, linen woven with coconut fibers, and horsehair.

Rhyme in Three Quarter Time. Artist: Vance Studley. A cast piece utilizing bamboo, flax, and jute fiber.

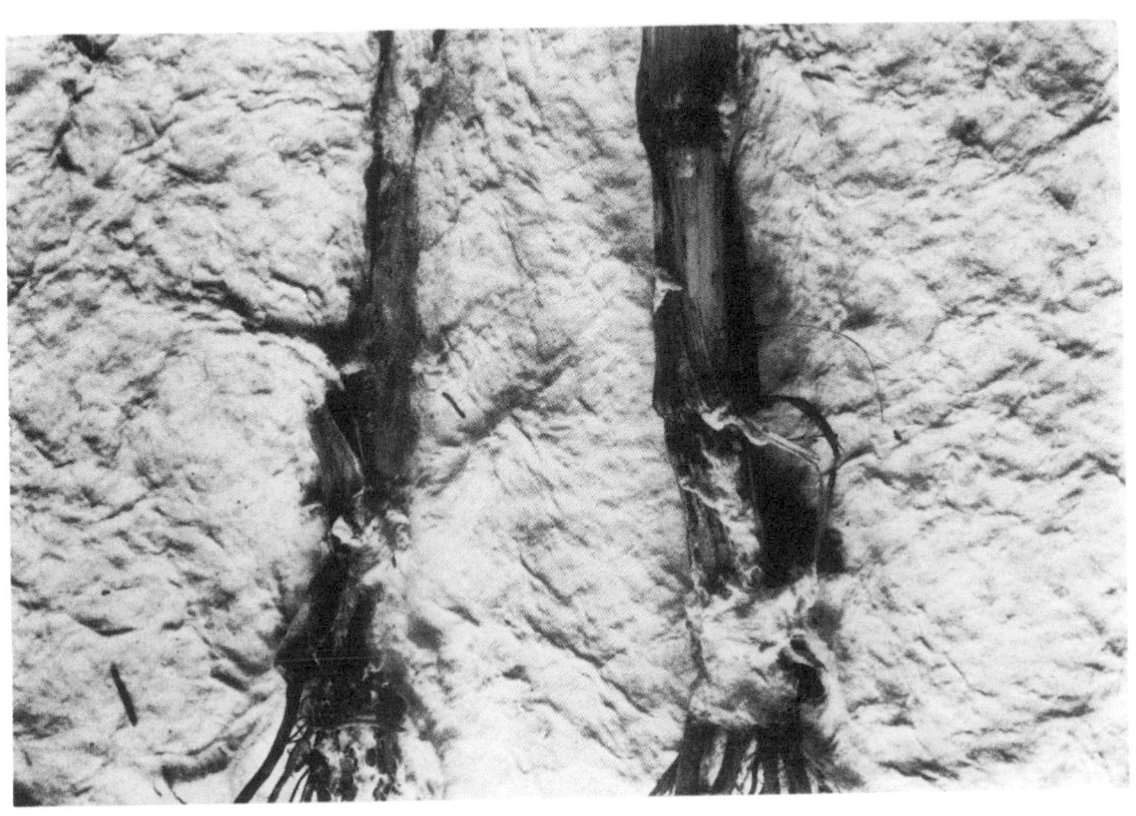

An enlarged view of *Rhyme.* Notice how the stalk of the plant appears to rise from the background suggesting a snow covered form.

A PORTFOLIO

Materials

Selected handmade papers
Cardboard
Bookcloth or booktape
mat knife
Ruler, metal-edged
White glue
Scissors
Tongue depressor
Stiff bristle brush (1 inch width)

Procedure

Many of the handmade papers you make will be delicate and will need to be protected. A portfolio for your work will provide necessary protection and will also serve as a carrying case and storage file. The portfolio can be small or large and its use will determine its dimensions. A very simple portfolio is made of flat cardboard joined together at its edges with booktape. The portfolio may be decorated inside and out with more unusual papers or fabric. Materials used should be durable and able to resist excessive handling. An enclosed portfolio, illustrated here, is a handsome and traditional folder that anyone can make by following the instructions and the diagram.

Cut the sheets of cardboard to the dimensions required as follows: A = C, B = D, E is half the width of A, and B is half the length of A. Leave a space of 1" between the cardboard to allow for folding. Once the boards have been cut and positioned, decide whether to glue paper or fabric to one or both sides of each board. Trim the excess paper allowing 1" to turn in. Glue the boards to the materials with white glue spread thinly and evenly over the entire surface. Burnish carefully with the tongue depressor. If excess glue is forced to the edges of the boards, simply wipe it off with a damp cloth.

Slit the corners, glue and turn down all edges. A patch may be necessary to cover all four corners. Make the patches triangular in shape, position and glue. As the glue begins to congeal burnish with the tongue depressor. In order to cover the hinges it is necessary to attach a thin lining of bookcloth or tape for additional reinforcement. Before inserting handmade sheets, check to be sure that all glue has thoroughly dried.

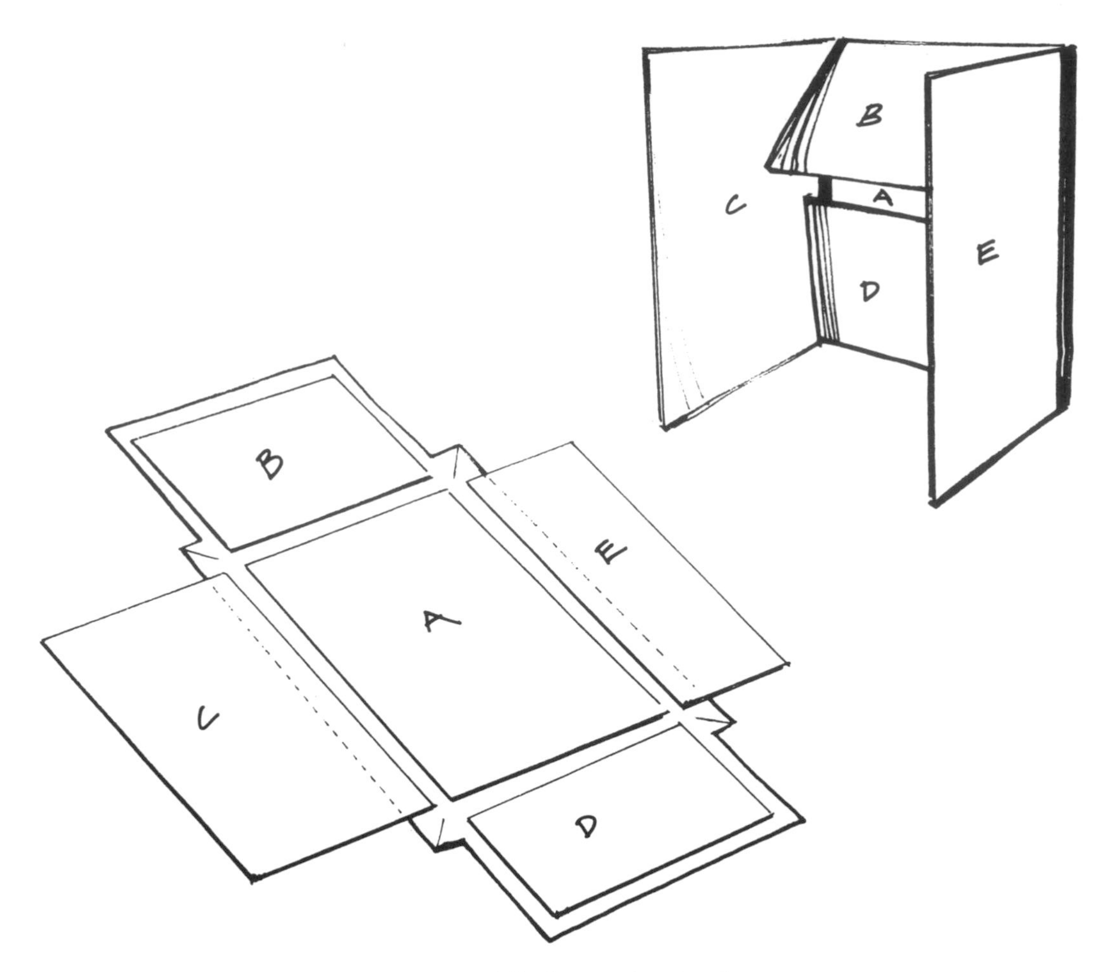

An enclosed portfolio for your handmade papers.

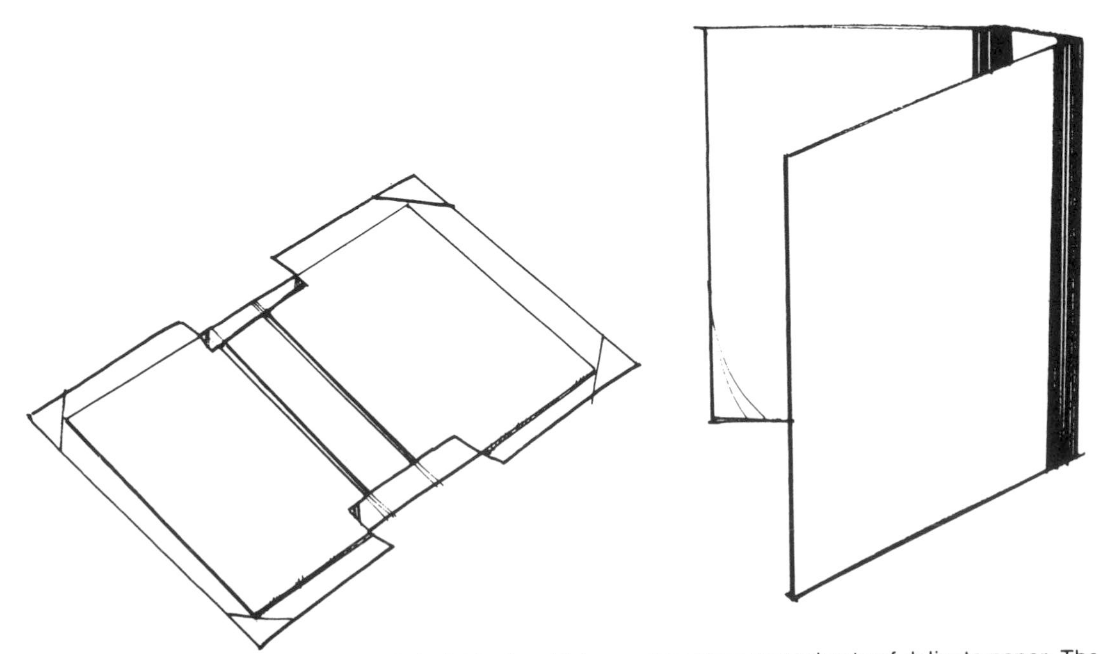

A portfolio folder in which you may store your sheets of delicate paper. The folder is made from double-weight art board, or binder's board, then bound along the edge of the spine with bookbinding tape available in most art supply stores.

PAPER FOR PRINTMAKING AND DRAWING

Materials

Several sheets of handmade paper of assorted sizes
Plate or textured objects to be used for printing (woodblock, linoleum, potato or onion slice, collagraph plate, etc.)
Plate of glass for inking
Inks (water or oil base)
Roller or press if available
Pen and ink
Charcoal pencils
Pastels
Pencils
Brayers

Procedure

Papers suitable for printing and drawing may come from almost any of the materials mentioned earlier. After several experiments you will find that paper made of highly texturous materials is too rough for drawing pictures requiring great detail but is suitable for making prints with use of a master plate. Collagraphs, woodblocks, etchings, potato prints, etc. can be used with sensitivity if they are combined with the right paper.

In printmaking, use papers that can withstand the pressure of the rolling pin or press. Ask yourself if the paper needs to be dampened before pulling the print. If so, has the paper been sufficiently dried and sized to accept the inked image under pressure? A deep line that has been gouged out of wood or metal requires paper that has the ability, under pressure, to conform to the depth of the cut. Some handmade papers are too brittle to do this even after lengthy soaking in a tub of water. These brittle papers are better suited for low relief transfer printing or drawing.

On the other hand, papers containing fiber that bends easily and willingly accepts ink and pressure and are highly useful to the printmaker. These papers can be made out of straw, artichoke, banana stalks, sized cattail, bamboo leaves, iris, and gladiola. Each of these plants has been used in making successful paper ideally matched to the printmaking medium.

If in doubt as to whether your paper is conducive to printmaking, experiment with a small sample. Dampen it with water, blot and rub ink onto its surface. Does the ink feather or discolor? How does it look to you when dry? Equally important, does the paper you have so dearly made with hours of effort, compliment the image to be placed on its surface? If the paper is to be embossed, does the relief show the image and paper in the best light? In time, you will find favorite papers for your visual ideas, but until then, try a few of the following techniques.

Sensitive prints can be made by inking such things as leaves, wood, string, wire, fish, cardboard, linoleum, potato sections, and body parts and then offsetting the image onto paper. All of this can be accomplished without the printmaker's press.

Voice of the Interior. Artist: Vance Studley. A print made on handmade paper of bamboo leaves.

Detail of *Voice* showing extreme embossment of the paper's surface achieved with the aid of an etching press.

Apply a thin veneer of ink with a rubber or gelatin brayer onto one of the above objects. Evenly spread the ink on the surface. Turn the inked object upside down and place it onto your handmade sheet. With a steady and direct motion, press the object into the paper as you would with a rubber stamp on stationery. Lift the object off the paper, and what remains is an impression, made in reverse, of the plate. Set the print to one side to dry.

If you want to make a print where the paper must bend to conform to an irregular surface such as a cardboard collage (collagraph) or if the image is too faint, using the prior process, then place the dampened handmade sheet on top of the already inked plate. Use a large metal spoon or rolling pin to burnish the backside of the paper gently into the valleys of the plate being careful not to force the paper beyond its limits. If you know your paper, as you will having been involved in its manufacture, you develop a sixth sense of what your paper can and cannot do. Handmade paper is a highly intimate material. Once the burnishing is completed, lift the paper off the plate in a peeling motion taking care not to blur the image. Your print will have an embossed, low-relief quality that makes it all the more intriguing.

Inking of the plate or nature form is not always necessary. A "blind" emboss print is made by side-stepping the inking process and proceeding to the pressing or rubbing stage. When the print is peeled back, the low-relief surface of the print can reveal the many subtle characteristics of your fine, handmade paper.

Drawing can be done on any piece of paper you can make. If the paper is unsized, then you must expect water media such as ink to feather and bleed. Many modes of drawing make this highly desirable. A soft, diffused line can imbue your work with great expressiveness and tonality. One of the most basic styles of using aqueous media on paper is to be found in watercolor. Blurred skies, oceanscapes, soft horizons, fluffy clouds, hazy backgrounds, and diaphanous landscapes can be drawn and painted on many unsized papers.

Sized paper can also accommodate these qualities which are better achieved with a soft brush. Sized sheets are ideal for pen and ink, charcoal, markers, pencil, and pastel.

Tribunal. Artist: Vance Studley. A print on flax paper achieved with the aid of a rolling pin and collage elements.

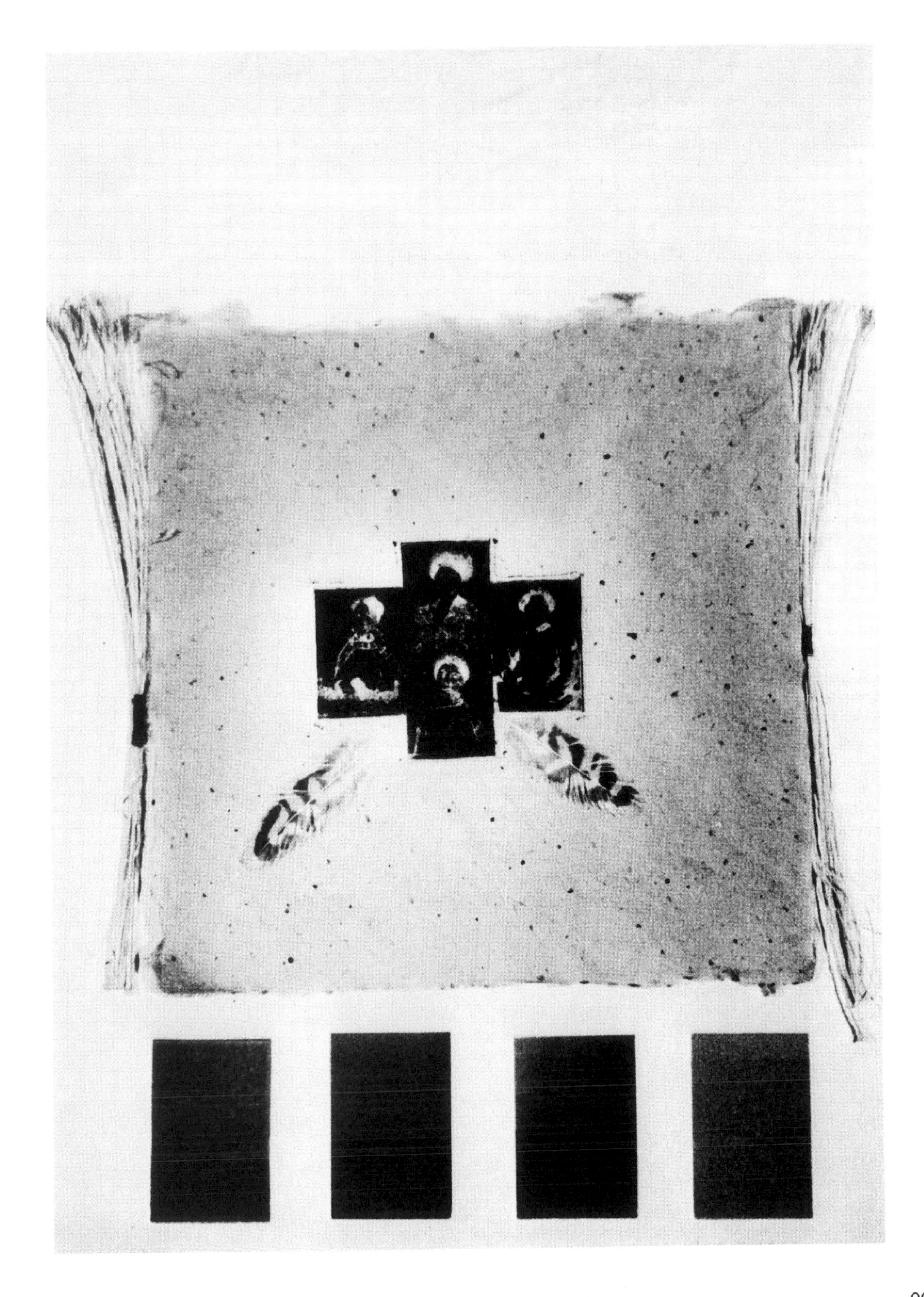

PAPER COLLAGE

Materials

Assorted pieces of large and small handmade paper
Gel medium (acrylic glue) or white glue
Spoon
Scissors
String
Sewing needle and multicolored thread
Tweezers
Glass, 9-by-12 inches

Procedure

An assortment of odds and ends of handmade paper can be transformed into provocative compositions of color and texture, achieving great beauty. Small jewel-like works of art can be made by arranging geometric and abstract shapes of paper on paper or thin wood veneer. By combining printed and handmade ephemera, the artist is able to structure new ideas of paper usage in a legitimate media. The collages of Paul Klee, Joseph Cornell, Kurt Schwitters, William Dole and Alberto Burri show an unpredictable fusion of fantasy and abstract elegance.

Modern artistic expression has given wide berth to the collage as a serious medium. Two-dimensional pattern in collage can suggest an illusionistic third dimension through careful placement and purposeful design. The collage can be surprising and stark or subdued and ponderous. Unlimited variety is possible with the use of carefully selected scraps of old and specially prepared papers, which, in part, have been treated with warm and cool washes of color.

Proceed by closely examining the surface of the paper and its space. Select paper for its unusual shape and size. Ideas will slowly present themselves to you as you progress.

Place a flat piece of paper, board, or wood veneer on a table. Next to this arrange several small boxes filled with assorted oddments of handmade paper. These may be grouped by color, weight, texture, and shape for ease of selection.

Select several small pieces of paper and begin to compose a simple design on your background piece stressing color, pattern, or other features. Extremely small pieces of paper can best be handled with tweezers. Try placing a straight edge of paper next to a ragged edge for contrast. Arrange color for impact and intensity.

When the beginnings of an idea take root, start to glue the backside of each piece of paper with gel medium or white glue. Position on the paper or board, then burnish the chip in place. Often glue will ooze out the sides if too much has been used. Simply take a damp cloth and gently wipe the excess glue away being careful not to disturb the composition.

Take time to closely scrutinize your growing composition for effectiveness of thought and visual form. Happy coincidences can be achieved by capriciously placing fugitive shapes and colors where ordinarily you might not have considered using them. The interaction of mottled papers juxtaposed with smooth, white, untextured ones can be visually exhilarating.

Once the composition has had a thorough going over, set it aside for the day by placing the glass sheet over the entire work. The lapsed time will allow the glue to congeal and affords you the chance to return the next morning for a fresh look at the previous day's efforts. A studied approach to the art of collage can make this an art of strategy between idea and form. By leaving a small area or boundary or uninterrupted space around the work the framing and matting process becomes simplified and does not infringe on the picture plane.

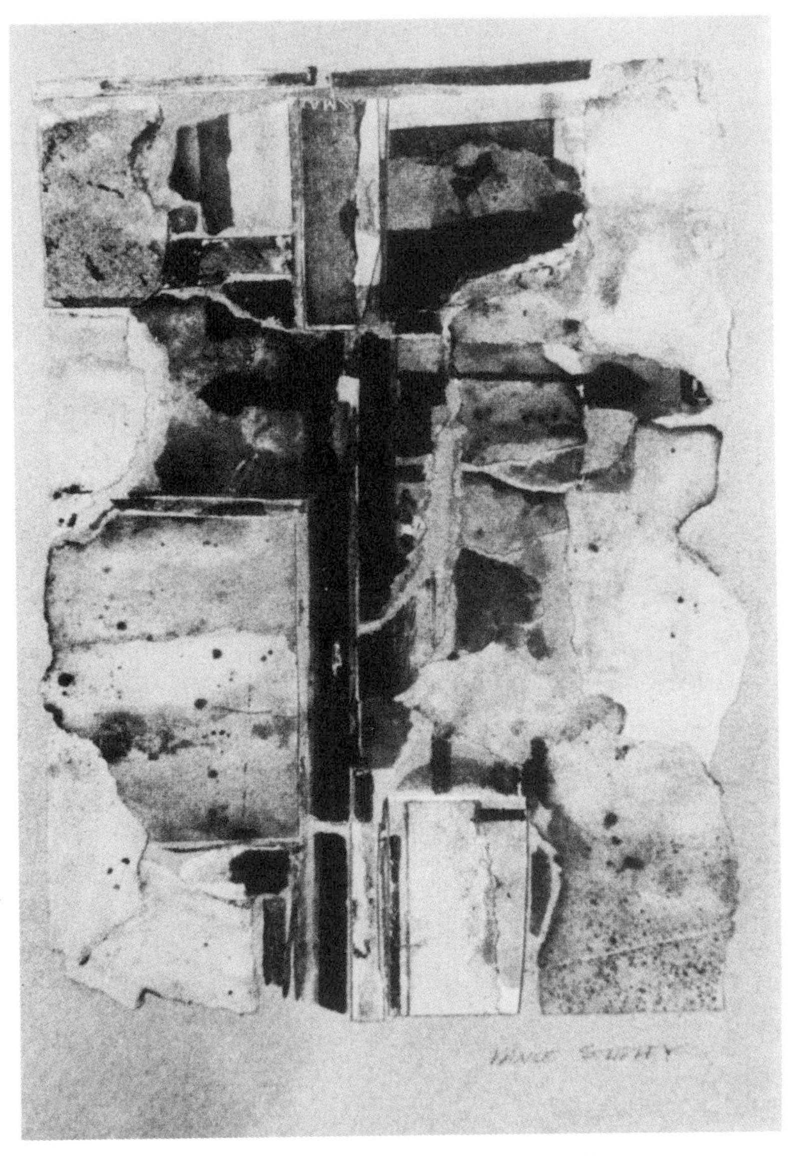

Totem. Artist: Vance Studley. A collage made up of small fragments of handmade paper tinted and toned with washes of Dr. Martin's Dyes. (Collection of Dr. Mark Simpson)

Certain Wind. Artist: Vance Studley. A collage of handmade paper of jute, carob, moss, and maple-leaf paper.

Untitled. Artist: Vance Studley. (Collection of Steve Wirth)

Roan (*Pages and Fuses*). Artist: Robert Rauschenberg. Handmade paper, screened images. (GEMINI G.E.L., Los Angeles, California)

Scow (*Pages and Fuses*). Artist: Robert Rauschenberg. Handmade paper; images screen printed on tissue paper and laminated to wet paper pulp. (GEMINI G.E.L., Los Angeles, California)

Marbled paper. Paper of this sort is made by immersing sheets of handmade or handmolded paper into a shallow tray into which has been poured small amounts of colorful lacquer paints. The marbling is achieved by gently fusing the colors with the aid of a toothpick as it trails its way through the paint. The paper is immersed for the briefest time then removed and allowed to dry for a few days. Marbled paper is ideal for collage owing to its colorful blend of paint and compatability with assorted bits of handmade paper.

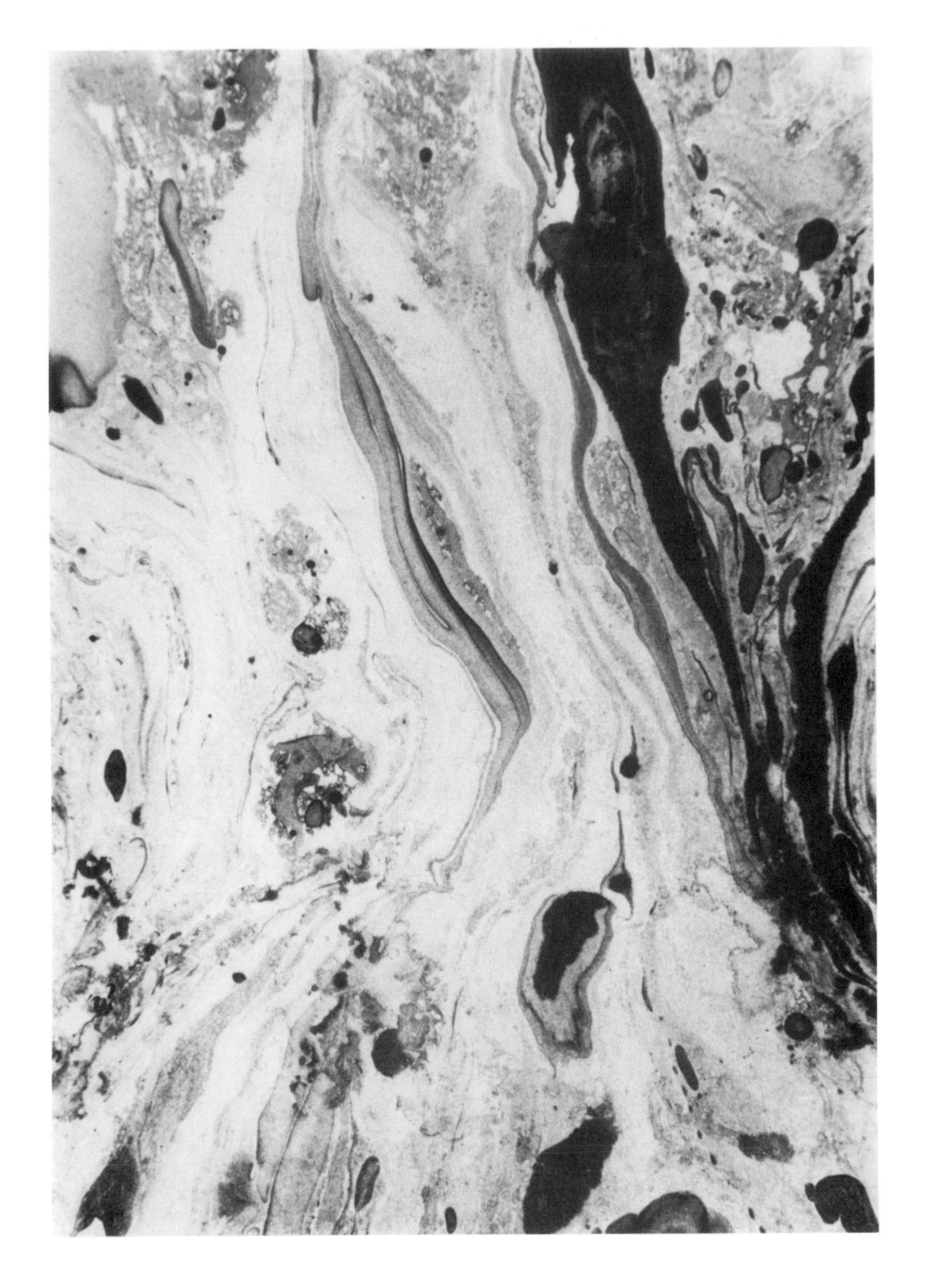

Glossary

Beater—a large mixer in which the pulp is mixed with the other ingredients of paper

Bleach—a chemical used to whiten paper pulp, a solution of chlorine or similar chemicals

Couch—a table, low in profile, used to support the formed sheets of paper. Also, to couch, meaning to transfer the sheet of paper from the mold to the felt blanket. This is accomplished by rolling the mold across the felt from one side to the other. Pronounced cōōch.

Deckle—a wooden frame that fits over a mold while a sheet of paper is being formed. The deckle prevents the paper pulp from overflowing the mold. This imparts to the edge of the paper the serrated or "feathered" edge, characteristic of untrimmed papers.

Emboss—a finish on paper achieved by means of a raised or depressed form in low relief so as to leave a visible surface or design on the paper

Felt—a square of absorbent material, usually woven wool or pressed felt, cut larger than the sheet to be formed, on which the formed sheet is to be couched. Also used to separate the newly cast sheet and to aid in drying.

Hollander—the paper beating machine that was invented in Holland

Laid—the ribbed appearance in some papers made by wires in the mold laid side by side rather than woven transversely

Macerate—to make parts of the substance soft or tender by soaking them in a liquid. In papermaking it is the process which reduces fiber to pulp and is usually accomplished in water with a beating, tearing, or shredding motion.

Mold—the rigid wooden frame on which the pulp is drawn from a tub or vat. The frame is covered with a wire screen or other suitable mesh.

Post—a stack of formed sheets with alternating felts. When the papermaker obtains a full stack, he presses it to remove excess water, usually with a clamping device, such as a book press.

Pulp—papermaking materials existing in a disintegrated, fibrous, wet, or dry state. Before it is disbursed in water, it is mixed, beaten, and diluted to a consistency suitable for fabrication into paper.

Sizing—a water resisting material which is added to paper. This can be done while the paper is in the pulp state or after the sheet has been dried.

Stuff—the material out of which paper is made. The stuff may be cotton, vegetable matter or wood cellulose.

Waterleaf—the newly formed wet, unsized sheet of paper

Watermark—a translucent marking made in paper while it is still wet for purposes of identification of the paper. The design of the watermark is sewn onto the screen of the mold with fine wire and can have many configurations.

Wove Mold—the screen on a mold which contains wire sewn transversely. Paper formed on a wove mold evinces no lines.

Bibliography

Ainsworth, J. H. *Paper the Fifth Wonder.* Wisconsin: 1959.

Butler, Frank. *The Story of Papermaking.* Chicago: Butler Paper Co., 1901.

Bayldon, Oliver. *The Paper Makers Craft.* Leicester, England: Twelve by Eight Press, 1965.

Clapperton, Robert. *Modern Paper Making.* London: Oxford University Press, 1941.

Cross, Charles. *A Textbook of Paper-Making.* London: 1900.

The Dun Emer Press. *The Nature and Making of Papyrus.* Barkston Ash, Yorkshire, England: Elmete Press, 1973.

Encyclopedia Brittanica. "History of Paper." 1975.

Fiore, Quentin. "Paper." *Industrial Design* 5 (1958) 32–60.

Green, J. Barcham. *Papermaking by Hand in 1967.* England: Maidstone, 1967.

von Hagen, Victor. *The Aztec and Maya Papermakers.* Locust Valley, New York: J. J. Augustin, Inc., 1944.

Higham, Robert. *A Handbook of Papermaking.* London: Oxford University Press, 1963.

Hollander, Annette. *Decorative Papers and Fabrics.* New York: Van Nostrand Reinhold Company, 1971.

Howell, Douglass. "The Future of Paper." *Print*, 14 (1960): 95–6.

Hunter, Dard. *My Life with Paper: An Autobiography.* New York: Alfred A. Knopf, Inc., 1958.

______. *Paper Making by Hand in America.* Chillicothe, Ohio: Mountain House Press, 1932.

______. *Papermaking in the Classroom.* New York: Manual Arts Press, 1931.

______. *Paper Making through Eighteen Centuries.* New York: Wm. Edward Rudge, 1930.

______. *Papermaking: The History and Technique of an Ancient Craft,* 2nd Ed. New York: Dover Publications, Inc., 1978.

Jugaku, B. *Paper Making by Hand in Japan.* Tokyo: Meiji-Shobu Ltd., 1959.

Kubiak, Richard. *The Handmade Paper Object.* Santa Barbara: The Santa Barbara Museum of Art, 1976.

Kuo-Chin, Liu. *Story of the Chinese Book.* Peking: Foreign Language Press, 1958.

Lehner, Ernst. *Symbols, Signs and Signets.* New York: Dover Publications, Inc., 1969.

Lewis, Naphtali. *Papyrus in Classical Antiquity.* London: Oxford University Press, 1974.

Mason, John. *Paper Making as an Artistic Craft.* London: Faber and Faber Ltd., 1959.

Morris, Henry. *Omnibus.* North Hills, Pennsylvania: Bird and Bull Press, 1967.

Munsell, Joel. *A Chronology of Paper and Paper-Making*, 4th Ed. Albany: 1870.

Ogawa, Hiroshi. *Forms of Paper.* New York: Van Nostrand Reinhold Company, 1971.

Reed, Ronald. *The Nature and Making of Parchment.* Barkston Ash, Yorkshire, England: Elmete Press, 1975.

"Robert Rauschenberg." *Print Collector's Newsletter.* 6 (1975): 137.

Tsien, Tsuen-Hsuin. *Written on Bamboo and Silk.* Chicago: University of Chicago Press, 1962.

U.S. Library of Congress. *Papermaking: Art and Craft.* Washington, D.C., 1968.

Voorn, Henk. *Old Ream Wrappers.* North Hills, Pennsylvania: Bird and Bull Press, 1969.

______. *De Schoolmeester.* Parnassusweg 126, Amsterdam, The Netherlands: Van Gelder Papier.

Wheelwright, Wm. Bond. *Practical Paper Technology.* 14 Hurlbut Street, Cambridge, Massachusetts: Wm. Bond Wheelwright, 1951.

Index

Note: Page numbers in italics indicate illustrative matter.

Africa, 16
alum for sizing, 35, 64, 76
amatl bark papers, 17, *24*
American Indians, 15
animal sizing, 19, 64
animal skins, writing materials from, 15–*16*, 17, 19
art, paper, used as form of, 26–28, *78–103*
Ashurbanipal, 12
Aztecs, 17, 24

bamboo for permanent writings, 14
bark of trees as writing material, 15
Baskerville, John, 43
Bayldon, Oliver, 63
beater, *see* electric beater
beating the pulp, 52, 59–63
bleaching ingredients, 32, 34, 35, *58*, 59
bleeding of colors, 82
blender, electric, 30, 32, *33*, 37, *40*, 52, 63, 80
blending the pulp, 30, 65
boiling the pulp, 30, 52, *56*–58, 59
bookbinder's nipping press, *75*
bookbinder's screw press, *74*, 84
Box #49, 89
brass for permanent writings, 12
bronze for permanent writings, 12
Burri, Alberto, 100
butcher wrap, *34*
butterflies, embedding of, in paper, 84

calender roll, *21*
calligraphy, 16, 17, 31
casting, *87*–90, *91*, *93*
cauldron for boiling raw material, 30, *56-57*
Central America, 15
Certain Wind, 102
Chaldeans, 12
China and Chinese:
  papermaking in, 17, *18*
  written communication before paper in, *11–12*, 14
Chlorox, 80
clay bricks and tablets, 12, *13*
cleanup, 36, 76, 77
codex, 14
Codex Monteleone, *24*
collage, paper, *99-102*
  marbled paper for, *104*
Cologne, Germany, 19
colored pulp:
  bleeding, 82
  for casting, 87
  for layering, 83
  paper sculptures with, 79
  process for, *64*
  unwanted, 80
copper for permanent writing, 12
Cornell, Joseph, 100
cuneiform, 12, *13*

Daumier, Honoré, 10
deckle, *see* mold and deckle
"deckle-edge," *42*
definitions, 106–107
Degas, Edgar, 9, 10
De Schoolmeester's paper, *23*
Dole, William, 100
Dr. Martin's Dyes, 64, *101*
draining process, *67-69*
drawing on handmade paper, 96–*99*
drying process, 31, *61*, 71, *72-75*, 77, 80, 82
  speeding up the, 79–80, 84, 90
dyeing ingredients, 34, 35, 64, *101*

Egypt and Egyptians, 18
  written communication before paper in, *12*, 16–17
electric beater, *67*
electric blender, 30, 32, *33*, 37, *40*, 52, *63*, 80
electric handmixer, 37, *40*
electric irons, *75*
electric juicers, 37
England, 18
equipment for papermaking, 10, *33*, *36*, 37–44
  see also materials; tools

Fabriano, Italy, 19

Feldman, Bella, *88*
felts, 30-31, 52, *69*, *72*, 84
  cleaning, after sizing, 76, 77
  post of, 31, *70*, 84
  separation from paper by machine, *25*
Formosa, 17
"foxing," 45, *56*
France, 18, 19
Franklin, Benjamin, 26
Fuller, Buckminster, 9

gelatin for sizing, 35, 64, *76*, 77
Germany, 19, *20*
glossary, 106-107
Goya, Francisco, 10
Greece, 15, 17
Gutenberg Bible, 16

Hals, Frans, 20
hammer and hollowed-out log, 52, *62*
hand beater, *67*
handmade paper, 10-11, *26*
  beating the pulp, 52, 59-63
  blending the pulp, 30, 65, 80
  boiling the pulp, 30, 52, *56*-58, 59
  collage with, *99-102*
  coloring the pulp, *64*, 83
  draining, *67-69*
  drying, 31, *61*, 71, 72-*75*, 77, 79-80, 82, 84, 90
  embedding objects in, 80-82
  forming the paper, 69-69
  history of, 16-26
  Japanese, 59, *61*, *73*, 82, *85*, *86*
  layering, 82-*86*
  materials, tools and equipment for, 32-51, 78, 80, 82, *83*, 87, 94, 96, 100, *102*, *104*
  portfolio for, 94-*95*
  pouring the pulp into a form (cast), *87*-90, *91*, *93*
  pressing, 52, 71, 72-*75*, 84
  for printmaking and drawing, *23*, 75, 96-*99*, *103*
  sculpture or relief using three-dimensional form, *78*-80, 90
  separation of, 71, 75, 79, 80
  sizing, 31, 64, *76*-77, 98
  steps for making, *21*-22, 30-31, 52-77
  washing the pulp, 30, 52, 58-59
handmixers, electric, 37, *40*
hanshi paper, *61*
Hilger, Charles, *89*
history:
  of papermaking, 16-26
  of written communication prior to paper, *11*-17
Holland, 20, *23*
Hollander beater, 20, *50-51*
hot plate, 37
Ho-ti, Emperor of China, 17
Hunter, Dard, 26
Hymenopterous (paper wasp), 52, *53*

iron, electric, for drying paper, *75*
Italy:
  early papermaking in, *19*
  watermarks and, 29
Ivy Mill, Pennsylvania, *25*, 26

Japan, 18
  handmade paper process in, 59, *61 73*
  layering of paper in, 82, *85*, *86*
  material used for paper in, 59, *61*, *86*
Java, 62
juicers, electric, 37

Kinds of paper, 10, 63
Klee, Paul, 9, 100
Korea, 18
kraft paper, *34*, 63

"laid" finish, *43*
Laubhan, Karen, *92*
layering of paper, 82-*86*
lead for permanent writings, 12
liquitex acrylic colors, 64
lye, 52, 57, *58*

machine-made paper, *6-7*, *23*, *25*
marbled paper, *104-105*
Marin, John, 9
materials
  for collage, 100, *102*, *104*
  for embedding objects in paper, 80
  for layering, 82, *83*
  for paper sculpture or relief, 78
  for a portfolio, 94
  for pouring pulp into a form (casting), 87
  for printmaking and drawing, 96
  suitable for papermaking, 6, 10, 17, *23*, 24, 30, 32, 34-35, *54-55*, 59, *61*, *86*
  see also equipment for papermaking; tools for papermaking
Mayan civilization, 17, 24
meat grinder, *39*, 40, *54*
metals for permanent writings, 12
Mexico, 15, 17, 24, 62
mold and deckle, 30, *41-49*, 52, *67-68*, *70*
  construction methods, *41*, 45-46, *47*
  transferring sheets to felts from, *69*
molds for casting process, 87-88
money, *7*
Montgolfier, Jean, 19
Morocco, 18
mottled paper, *65*
Mountain House Press, 26

Netherlands, 20, *23*
Nuremberg, Germany, 19, *20*
Nuremberg Chronicle (Schedel), 19-*20*

oracle bones, *11*-12
Otomi Indians, 24, 62

ox bones, *11*

Pacific Islands, 62
palmyra leaves, writing on, *14*
paper currency, *7*
paper toweling, *7*
paper wasp, 52, *53*
papyrus, 16–*17*
parchment, 15–*16*, 19
Paris, Harold, *89*
Pennsylvania, paper mills in, 24–*25*, 26
Pergamum, Mysia, 15
Picasso, Pablo, 9
pictograms, 11
portfolio, 94–*95*
pot for boiling raw material, 30, *56–57*
pressed finish, 31
pressing, 52, 71
 devices for, *72–75*, 84
 of layered paper, 84
Primordial Dawning, *90*
printing and printmaking:
 on handmade paper, *23*, 75, 96–*99*, *103*
 on machine-made paper, *23*
Prophesies of My Future Self 1976-1996, *89*
pulp, 6, 30, 52
 beating, 52, 59–63
 blending, 30, 65, 80
 boiling, 30, 52, *56*–58, 59
 coloring, *64*, 83
 embedding objects in, 80–82
 layering, 82–*86*
 machines for macerating, 20, *21*, 30, 32, *33*, 37, *40*, *50–51*, 63
 paper, as an art form, 26–28, *78*–103
 pouring, into a form (cast), *87*–90, *91*, *93*
 pouring, over three-dimensional form, *78*–80, *90*
 washing, 52, 58–59

Rauschenberg, Robert, *103*
relief, paper, 78–80, *90*
Rembrandt van Rijn, 10, 20
rice paper, 17
Rittenhouse, Jacob, 25
Rittenhouse, Klaus, *24*
Rittenhouse, Nicolas, 25
Rittenhouse, William, 24–25
Roan, *103*
Robert, Nicholas-Louis, *6*
rolling pin, 74
Romans, writing materials used by, 14, 15
rosin, powered, for sizing, 64, 77

Schedel, Hartmann, 19–*20*
Schwitters, Kurt, 100
scissoring the raw material, 54–*55*
Scow, *103*
sculpture, paper, *78*–80
separating the sheets, 71, 75, 79, 80
shoji panels, 28
silk waste for paper, 17
sizing, 31, 64, *76–77*, 98
sizing ingredients, 34, 35, *64*, *76*, 77
slaked lime, 52
South America, 15, *24*
Spain, 18
spices, 84
stamping machine, *62*
stone, writing on, *12*
Stromer, Ulman, 19, *20*
Studley, Margo, *91*
Studley, Vance, *90*, *93*, *97*, *99*, *101–102*
sumi painting, 17

tapa cloth, *62*
texture of paper, 6, 10, *26*, *71*, *73*, *97*
 drawing and, 96
 layering, and, 82–83, 84
three-dimensional forms, pouring pulp over, *78*–80
Three Quarter Time, *93*
"throwing off the wave," 30, *67*, 68
tools for papermaking, 10, *36*
 see also equipment for papermaking; materials
Totem, 101
trees, bark of, as writing material, 15
Tribunal, *99*
trip-hammer, *62*, 63
Troyes, France, 18
Ts'ai Lun, 17, *18*
tubs, busboy, *66*, *70*

United States, papermaking in, history of, *24*–26

Valencia, Spain, 18
Van Gelder Paper Company, 23
vat, 30, 52, 59, *66*, *70*
vegetable shredder, *55*
vegetable spray, 79, 87, 88, 90
vegetation, writing on, *14*, *16*–17
vellum, 15–*16*, 19
Vermeer, Jan, 20
Voice of the Interior, *97*

washing the pulp, 52, 58–59
watercolor on handmade paper, 75, 98
waterleaf, defined, 72
watermarks, *24*, *28–29*, *43*
Willcox, Thomas, *25*
wood for permanent writings, 14
"wove" finish, 42–43
written communication, materials used for, before paper, *11*–17